APOLLO PROGRAM FLIGHT SUMMARY

Mission Designation	Launch Date	Mission Description
Apollo Saturn IB (AS-201)	Feb 26, 1966	Unmanned, suborbital, space vehicle development flight. Demonstrated space vehicle compatibility and structural integrity; spacecraft heat shield qualification for earth orbital reentry speeds.
Apollo Saturn IB (AS-203)	July 5, 1966	Unmanned, orbital, launch vehicle development flight. Demonstrated second stage restart and cryogenic propellants storage at zero g conditions. Liquid hydrogen pressure test.
Apollo Saturn IB (AS-202)	Aug 25, 1966	Unmanned, suborbital, space vehicle development flight. Demonstrated structural integrity and compatibility, spacecraft heat shield performance.
Apollo 4 (AS-501)	Nov 9, 1967	First Apollo Saturn V flight, unmanned, earth orbital to 11,234 miles apogee, space vehicle development flight. Demonstrated Saturn V rocket performance and Apollo spacecraft heat shield for lunar mission reentry speeds.
Apollo 5 (AS-204 LM)	Jan 22, 1968	First Apollo lunar module flight on Saturn IB, unmanned, earth orbital. Demonstrated spacecraft systems performance, ascent and descent stage propulsion firings and restart, and staging.
Apollo 6 (AS-502)	April 4, 1968	Second flight of Saturn V, unmanned, earth orbital, launch vehicle development flight. Demonstrated Saturn V rocket performance and Apollo spacecraft subsystems and heat shield performance.

(Continued on back cover)

MANNED SPACE FLIGHT NETWORK

POSTMISSION REPORT

on the

AS-506 (APOLLO 11) MISSION

February 1970

Prepared by

Operations Analysis Section
Systems Operations Branch
National Aeronautics and Space
Administration

Support Group
Systems Operations Section
Bendix Field Engineering Corp.

Submitted by

Concurred with

D.J. Graham, Head
Operations Analysis Section
Systems Operations Branch

Carl O. Roberts, Head
Systems Operations Branch
Manned Flight Operations Division

Approved by H. William Wood, Chief
Manned Flight Operations Division

GODDARD SPACE FLIGHT CENTER

Greenbelt, Maryland

Frontispiece. MARS 210-foot Antenna at Goldstone, California

ACKNOWLEDGEMENTS

This is to gratefully acknowledge that information used in this report came from persons throughout the MSFN, including particularly the personnel assigned as members of the GSFC Network Support Team during the AS-506 Mission.

Source materials used in this report include:

- The Network Operations Directive for NASA Manned Space Flight Operations as supplemented for AS-506.

- Teletype status messages transmitted by MSFN stations.

- Postlaunch Instrumentation Message data sheets prepared by systems personnel at MSFN stations.

- Station M&O Mission Reports.

- Network Controller's Mission Report for AS-506, MSC.

- Network Operations Manager's Report for AS-506.

- Appendix A to this document, provided by the NASA Communications Network.

- Radar tracking summary messages from individual stations.

- Summary of tracking information for AS-506, prepared by Flight Support Division, MSC.

- ARIA Controller's Report, AS-506 Mission, prepared by the ARIA Controller and the AOCC Support Team.

- Quick Look Data Analysis, ARIA support of AS-506, prepared by Headquarters, Air Force Eastern Test Range, ARIA Task Force, Patrick AFB, Florida.

- MSFN Metric Tracking Performance Report for AS-506, prepared by the Manned Flight Planning and Analysis Division, GSFC.

CONTENTS

Section	Page
INTRODUCTION	xv
1. MISSION AND SUPPORT SUMMARY	1-1
1.1 General	1-1
1.2 NRT Period (TSI 001 to ISI 001)	1-3
1.2.1 General	1-3
1.2.2 Tracking	1-4
1.2.3 Telemetry	1-5
1.2.4 Command	1-6
1.2.5 Air-to-Ground Communications	1-7
1.3 Prelaunch Period (ISI 001 to Launch)	1-7
1.3.1 General	1-7
1.3.2 Tracking	1-7
1.3.3 Telemetry	1-8
1.3.4 Command	1-9
1.3.5 Air-to-Ground Communications	1-9
1.4 Launch Phase	1-10
1.5 Earth Orbit Phase	1-10
1.6 Translunar Coast Phase, Including TLI	1-10
1.7 Lunar Phase	1-13
1.8 Transearth Coast Phase, including TEI	1-17
1.9 Reentry and Splashdown	1-17
2. NETWORK PERFORMANCE AND STATISTICAL ANALYSIS	2-1
2.1 General	2-1
2.1.1 Explanation of Statistics	2-1
2.1.2 Definition of Statistical Terms	2-1
2.2 Tracking	2-1
2.2.1 USB	2-1
2.2.2 C-band	2-21
2.2.3 VHF Tracking	2-21
2.3 Telemetry	2-24
2.4 Command	2-27
2.4.1 General	2-27
2.4.2 Command Activity	2-27
2.4.3 USB Command	2-27
2.4.4 UHF Command	2-29
2.5 Air-to-Ground Communications	2-29

v

Section	Page
3. STATION PERFORMANCE	3-1
3.1 General	3-1
3.1.1 General Problems	3-1
3.1.2 USB	3-2
3.1.3 Telemetry	3-3
3.1.4 Command	3-4
3.1.5 Air-to-Ground Communications	3-4
3.2 SPAN (Figure 3-1)	3-5
3.2.1 General	3-5
3.2.2 Mission Activity	3-6
3.3 ARIA (Figures 3-2 through 3-4)	3-6
3.3.1 General	3-6
3.3.2 Mission Performance	3-9
3.4 Antigua (ANG) (Figures 3-5 through 3-9)	3-9
3.4.1 General (Figures 3-5 through 3-7)	3-9
3.4.2 Tracking	3-11
3.4.3 Telemetry (Figure 3-9)	3-14
3.4.4 Command	3-15
3.4.5 Air-to-Ground Communications	3-15
3.5 Ascension (ACN) (Figures 3-10 through 3-14)	3-15
3.5.1 General (Figures 3-10 through 3-12)	3-15
3.5.2 Tracking	3-15
3.5.3 Telemetry (Figure 3-14)	3-16
3.5.4 Command	3-21
3.5.5 Air-to-Ground Communications	3-21
3.6 Bermuda (BDA) (Figures 3-15 through 3-21)	3-22
3.6.1 General (Figures 3-15 through 3-17)	3-22
3.6.2 Tracking	3-24
3.6.3 Telemetry (Figure 3-21)	3-28
3.6.4 Command	3-28
3.6.5 Air-to-Ground Communications	3-31
3.7 California (CAL) (Figures 3-22 through 3-25)	3-31
3.7.1 General (Figures 3-22 through 3-24)	3-31
3.7.2 Tracking	3-32
3.7.3 Telemetry	3-33
3.7.4 Command	3-33
3.7.5 Air-to-Ground Communications	3-33

Section	Page

3.8 Canary Island (CYI) (Figures 3-26 through 3-30)............ 3-33

    3.8.1 General (Figures 3-26 through 3-28) 3-33

    3.8.2 Tracking 3-34

    3.8.3 Telemetry (Figure 3-30) 3-38

    3.8.4 Command 3-38

    3.8.5 Air-to-Ground Communications 3-38

3.9 Carnarvon (CRO) (Figures 3-31 through 3-36) 3-40

    3.9.1 General (Figures 3-31 through 3-33) 3-40

    3.9.2 Tracking 3-40

    3.9.3 Telemetry (Figure 3-36) 3-44

    3.9.4 Command 3-46

    3.9.5 Air-to-Ground Communications 3-47

3.10 Goldstone/Goldstone Wing (GDS/GDSX) (Figures 3-37 through 3-42)............................. 3-47

    3.10.1 General (Figures 3-37 through 3-39) 3-47

    3.10.2 Tracking........................... 3-49

    3.10.3 Telemetry (Figure 3-42) 3-53

    3.10.4 Command 3-58

    3.10.5 Air-to-Ground Communications 3-58

3.11 Grand Bahama Island (GBM) (Figures 3-43 through 3-47) 3-59

    3.11.1 General (Figures 3-43 through 3-45) 3-59

    3.11.2 Tracking 3-59

    3.11.3 Telemetry (Figure 3-47) 3-60

    3.11.4 Command 3-63

    3.11.5 Air-to-Ground Communications 3-63

3.12 Guam (GWM) (Figures 3-48 through 3-52) 3-63

    3.12.1 General (Figures 3-48 through 3-50) 3-63

    3.12.2 Tracking 3-63

    3.12.3 Telemetry (Figure 3-52) 3-67

    3.12.4 Command 3-68

    3.12.5 Air-to-Ground Communications 3-68

3.13 Guaymas (GYM) (Figures 3-53 through 3-57) 3-68

    3.13.1 General (Figures 3-53 through 3-55) 3-68

    3.13.2 Tracking 3-69

    3.13.3 Telemetry (Figure 3-57) 3-70

    3.13.4 Command 3-75

    3.13.5 Air-to-Ground Communications 3-75

Section	Page

3.14 Hawaii (HAW) (Figures 3-58 through 3-63)   3-75

    3.14.1 General (Figures 3-58 through 3-60)   3-75

    3.14.2 Tracking   3-76

    3.14.3 Telemetry (Figure 3-63)   3-77

    3.14.4 Command   3-80

    3.14.5 Air-to-Ground Communications   3-81

3.15 Honeysuckle Creek/Honeysuckle Creek Wing (HSK/HSKX)
    (Figures 3-64 through 3-69)   3-81

    3.15.1 General (Figures 3-64 through 3-69)   3-81

    3.15.2 Tracking   3-83

    3.15.3 Telemetry (Figure 3-69)   3-87

    3.15.4 Command   3-88

    3.15.5 Air-to-Ground Communications   3-90

3.16 Huntsville (HTV) (Figures 3-70 through 3-75)   3-90

    3.16.1 General (Figures 3-70 through 3-72)   3-90

    3.16.2 Tracking   3-91

    3.16.3 Telemetry (Figure 3-75)   3-91

    3.16.4 Command   3-92

    3.16.5 Air-to-Ground Communications   3-93

3.17 Madrid/Madrid Wing (MAD/MADX) (Figures 3-76 through
    3-81)   3-93

    3.17.1 General (Figures 3-76 through 3-78)   3-93

    3.17.2 Tracking   3-95

    3.17.3 Telemetry (Figure 3-81)   3-98

    3.17.4 Command   3-100

    3.17.5 Air-to-Ground Communications   3-102

3.18 Mercury (MER) (Figures 3-82 through 3-87)   3-103

    3.18.1 General (Figures 3-82 through 3-84)   3-103

    3.18.2 Tracking   3-104

    3.18.3 Telemetry (Figure 3-87)   3-106

    3.18.4 Command   3-107

    3.18.5 Air-to-Ground Communications   3-107

3.19 Merritt Island (MIL) (Figures 3-88 through 3-93)   3-107

    3.19.1 General (Figures 3-88 through 3-90)   3-107

    3.19.2 Tracking   3-108

    3.19.3 Telemetry (Figures 3-92 and 3-93)   3-112

    3.19.4 Command   3-115

    3.19.5 Air-to-Ground Communications   3-115

Section	Page
3.20 Redstone (RED) (Figures 3-94 through 3-99).	3-116
3.20.1 General (Figures 3-94 through 3-96)	3-116
3.20.2 Tracking	3-117
3.20.3 Telemetry (Figure 3-99)	3-118
3.20.4 Command	3-121
3.20.5 Air-to-Ground Communications	3-121
3.21 Tananarive (TAN) (Figures 3-100 through 3-104).	3-121
3.21.1 General (Figures 3-100 through 3-102)	3-121
3.21.2 Tracking.	3-121
3.21.3 Telemetry (Figure 3-104)	3-123
3.21.4 Command	3-123
3.21.5 Air-to-Ground Communications	3-123
3.22 Texas (TEX) (Figures 3-105 through 3-109)	3-123
3.22.1 General (Figures 3-105 through 3-107)	3-123
3.22.2 Tracking.	3-123
3.22.3 Telemetry (Figure 3-109)	3-124
3.22.4 Command	3-126
3.22.5 Air-to-Ground Communications	3-126
3.23 Vanguard (VAN) (Figures 3-110 through 3-115)	3-128
3.23.1 General (Figures 3-110 through 3-112)	3-128
3.23.2 Tracking.	3-128
3.23.3 Telemetry (Figure 3-115).	3-128
3.23.4 Command	3-129
3.23.5 Air-to-Ground Communications	3-129
APPENDIX A -- NASCOM NETWORK PERFORMANCE	A-1
APPENDIX B -- LIST OF ABBREVIATIONS AND ACRONYMS	B-1
DISTRIBUTION LIST	DL-1

ILLUSTRATIONS

Figure		Page
2-1	Actual S-band Track to Predicted View	2-7
2-2	Percentage of Valid S-band Tracking to Total Track	2-9
2-3	Percentage of Actual Track to Predicted View Time (USB) -- by Mission	2-14
2-4	Percentage of Valid Track to Total Track (USB) -- by Mission . . .	2-14
2-5	Distribution of Percent of Actual Track to Predicted View Data Points for AS-506 and Previous Lunar Missions	2-15

ix

Figure		Page
2-6	Distribution of Percent of Valid to Total Track Data Points for AS-506 and AS-505	2-16
2-7	Percent of Actual Track to Predicted View Time -- by Station	2-17
2-8	Percent of Valid Track to Total View Time -- by Station	2-19
2-9	Percentage of C-band Actual Track to Predicted View	2-22
2-10	Percentage of Valid C-band Track During Total Tracking Time ...	2-22
2-11	Comparison of Percentage of Actual Track to Predicted View (C-band) -- by Mission	2-23
2-12	Comparison of Percentage of Valid Track to Total Track (C-band) -- by Mission	2-24
3-1	A Typical SPAN Station	3-5
3-2	ARIA in Flight	3-6
3-3	ARIA TLI Support Positions	3-8
3-4	ARIA Reentry Support Positions	3-9
3-5	ANG Tracking Station.................................	3-11
3-6	ANG Mission Support	3-11
3-7	ANG Support Periods	3-12
3-8	ANG USB Tracking Coverage	3-12
3-9	ANG Telemetry Coverage	3-14
3-10	ACN Tracking Station	3-16
3-11	ACN Mission Support	3-16
3-12	ACN Support Periods	3-17
3-13	ACN USB Tracking Coverage	3-18
3-14	ACN Telemetry Coverage	3-20
3-15	BDA Tracking Station.................................	3-22
3-16	BDA Mission Support	3-23
3-17	BDA Support Periods	3-23
3-18	BDA USB Tracking Coverage	3-25
3-19	BDA FPQ-6 Tracking Coverage	3-27
3-20	BDA FPS-16 Tracking Coverage	3-28
3-21	BDA Telemetry Coverage	3-29
3-22	CAL Tracking Station.................................	3-31
3-23	CAL Mission Support	3-32
3-24	CAL Support Periods	3-32
3-25	CAL TPQ-18 Tracking Coverage	3-33
3-26	CYI Tracking Station	3-34
3-27	CYI Mission Support	3-35
3-28	CYI Support Periods	3-35
3-29	CYI USB Tracking Coverage	3-36

Figure		Page
3-30	CYI Telemetry Coverage	3-38
3-31	CRO Tracking Station	3-40
3-32	CRO Mission Support	3-41
3-33	CRO Support Periods	3-42
3-34	CRO USB Tracking Coverage	3-42
3-35	CRO FPQ-6 Tracking Coverage	3-44
3-36	CRO Telemetry Coverage	3-45
3-37	GDS/GDSX Tracking Station	3-47
3-38	GDS/GDSX Mission Support	3-48
3-39	GDS/GDSX Support Periods	3-48
3-40	GDS USB Tracking Coverage	3-50
3-41	GDSX USB Tracking Coverage	3-53
3-42	GDS/GDSX Telemetry Coverage	3-56
3-43	GBM Tracking Station	3-60
3-44	GBM Mission Support	3-60
3-45	GBM Support Periods	3-61
3-46	GBM USB Tracking Coverage	3-61
3-47	GBM Telemetry Coverage	3-62
3-48	GWM Tracking Station	3-64
3-49	GWM Mission Support	3-64
3-50	GWM Support Periods	3-65
3-51	GWM USB Tracking Coverage	3-65
3-52	GWM Telemetry Coverage	3-67
3-53	GYM Tracking Station	3-69
3-54	GYM Mission Support	3-69
3-55	GYM Support Periods	3-70
3-56	GYM USB Tracking Coverage	3-71
3-57	GYM Telemetry Coverage	3-73
3-58	HAW Tracking Station	3-75
3-59	HAW Mission Support	3-76
3-60	HAW Support Periods	3-76
3-61	HAW USB Tracking Coverage	3-77
3-62	HAW FPS-16 Tracking Coverage	3-79
3-63	HAW Telemetry Coverage	3-79
3-64	HSK/HSKX Tracking Station	3-81
3-65	HSK/HSKX Mission Support	3-82
3-66	HSK/HSKX Support Periods	3-82
3-67	HSK USB Tracking Coverage	3-84

Figure		Page
3-68	HSKX USB Tracking Coverage	3-86
3-69	HSK/HSKX Telemetry Coverage	3-88
3-70	USNS Huntsville	3-91
3-71	HTV Mission Support	3-91
3-72	HTV Support Period	3-92
3-73	HTV USB Tracking Coverage, Reentry -- CSM	3-92
3-74	HTV CAPRI Tracking Coverage, Reentry -- CM	3-93
3-75	HTV Telemetry Coverage, Reentry -- CM	3-93
3-76	MAD/MADX Tracking Station	3-94
3-77	MAD/MADX Mission Support	3-94
3-78	MAD/MADX Support Periods	3-95
3-79	MAD USB Tracking Coverage	3-96
3-80	MADX USB Tracking Coverage	3-99
3-81	MAD/MADX Telemetry Coverage	3-101
3-82	USNS Mercury	3-103
3-83	MER Mission Support	3-104
3-84	MER Support Periods	3-104
3-85	MER USB Tracking Coverage	3-105
3-86	MER FPS-16 Tracking Coverage, EO -- IU	3-106
3-87	MER Telemetry Coverage, EO	3-106
3-88	MIL Tracking Station	3-108
3-89	MIL Mission Support	3-108
3-90	MIL Support Periods	3-109
3-91	MIL USB Tracking Coverage	3-109
3-92	MIL USB CSM PM Downlink Telemetry Configuration --Launch	3-113
3-93	MIL Telemetry Coverage	3-114
3-94	USNS Redstone	3-116
3-95	RED Mission Support	3-117
3-96	RED Support Periods	3-117
3-97	RED USB Tracking Coverage	3-118
3-98	RED FPS-16 Tracking Coverage, EO--IU	3-119
3-99	RED Telemetry Coverage	3-120
3-100	TAN Tracking Station	3-121
3-101	TAN Mission Support	3-121
3-102	TAN Support Periods	3-122
3-103	TAN CAPRI Radar Tracking Coverage, EO -- IU	3-122
3-104	TAN Telemetry Coverage, EO	3-123
3-105	TEX Tracking Station	3-124

Figure		Page
3-106	TEX Mission Support	3-124
3-107	TEX Support Periods	3-125
3-108	TEX USB Tracking Coverage	3-125
3-109	TEX Telemetry Coverage	3-127
3-110	USNS Vanguard	3-129
3-111	VAN Mission Support	3-129
3-112	VAN Support Periods	3-130
3-113	VAN USB Tracking Coverage	3-130
3-114	VAN FPS-16 Tracking Coverage, EO -- IU	3-131
3-115	VAN Telemetry Coverage, EO	3-132

TABLES

Number		Page
2-1	Mission Data Received -- by Station	2-2
2-2	Points Deleted	2-5
2-3	Statistical Summary of the Network USB Tracking Performance During AS-506	2-6
2-4	Statistical Summary of USB Tracking Performance by Station During AS-506	2-10
2-5	Tracking Performance During Lunar Stay	2-12
2-6	Statistical Summary of USB Tracking Performance During Lunar Mission	2-12
2-7	Statistical USB Performance During AS-503 and AS-506 -- by Phase	2-13
2-8	C-band Statistical Summary for AS-506	2-21
2-9	Statistical Summary of Radar Systems Performance During Lunar Mission	2-23
2-10	Statistical Summary of VHF Tracking Performance During AS-506	2-24
2-11	Percentage of Dropout Time to Total Lock Time -- by Mission Phases for USB Telemetry	2-25
2-12	Percentage of Dropout Time to Total Lock Time -- by Vehicle for USB Telemetry	2-25
2-13	Percentage of Dropout Time to Total Lock Time -- by Tracking System	2-25
2-14	Percentage of Dropout Time to Total Lock Time -- by Type of Station	2-26
2-15	MSFN Loading and Command Activity	2-28
2-16	Quality and Configuration of Spacecraft Communications	2-30
3-1	ARIA Support Summary	3-10

INTRODUCTION

This report reviews and evaluates the performance of the Manned Space Flight Network during the AS-506 Apollo Mission. This is accomplished by using inputs from the Manned Space Flight Network and documents such as the Network Operations Manager's, Network Controller's, and Apollo Range Instrumentation Aircraft Controller's Reports, Status Messages, and Maintenance and Operations Postmission Reports. These inputs have been analyzed and the resulting information, showing the performance of the entire network and of each Manned Space Flight Network station, is presented in this report.

Section 1 is a profile of the mission, tracing the flight from launch to splashdown and encompassing Manned Space Flight Network support. It is not intended to go too deeply into detail, but rather to give an overall picture.

Section 2 presents the statistical analysis, in tabular and graphical form, of the separate functional systems: tracking, telemetry, command, and communications.

Section 3 presents the coverage of each tracking station, describing in detail individual system support. The information is presented in the same order as in Section 2. The activities of the Solar Particle Alert Network and of the Apollo Range Instrumentation Aircraft are also included in this section. System constraints and limitations, as well as problems that were encountered, are shown. If a paragraph heading is not followed by any write-up, this means that at that station the named system supported as it was supposed to do and had no operator or equipment problems.

Appendix A is the NASA Communications Network mission report.

Appendix B contains abbreviations and acronyms.

1. MISSION AND SUPPORT SUMMARY

1.1 GENERAL

On July 16, 1969 the AS-506 Saturn V launch vehicle lifted off from Complex 39A, Cape Kennedy at 13:32:00 GMT. Aboard this mighty vehicle were Neil Armstrong, CDR; Edwin Aldrin, LMP; and Michael Collins, CMP. This was the start of a mission to fulfill an objective proposed by the late President John F. Kennedy:

> "I believe that this Nation should commit itself to achieving the goal, before this decade is out, of landing a man on the moon and returning him safely to earth."
>
> (May 25, 1961)

The flight was nominal from launch through LM DOI and, during TLC, the trajectory of the CSM/LM was so precise that midcourse corrections 1, 3, and 4 were not required. There were anxious moments at MCC after the LM crew initiated PDI and were under way for touchdown on the lunar surface. At 40,000 feet above the surface, the LM's guidance and navigation system began to flash an alarm warning that the computer was overloaded and was rejecting demands for additional data. When MCC was queried by the crew concerning this difficulty, they were told that the mission status was still "go." When the LM had descended to 27,000 feet above the lunar surface, the alarm again flashed to indicate the presence of radar difficulty. MCC again told the crew that the mission was "go." The data overload, according to MCC, occurred because of simultaneous computer inputs from the landing and rendezvous radars. The landing radar was supplying velocity and altitude data while the rendezvous radar was attempting to input the LM's position relative to that of the orbiting CSM. An alert member of the ground control team at MCC recognized what was causing the problem and through Flight Control directed the LM crew to cease questioning the computer on the operation of the landing radar. Telemetry from the LM was indicating to MCC that the operation of this radar was nominal. With the computer overload now somewhat relieved, the computer, with an assist from the LM crew (semi-automatic mode), brought the craft safely over a large boulder-filled crater to a successful landing on the lunar surface, approximately 4 miles downrange from the predicted landing site. The ingenuity of man, at MCC and within the LM, prevented an almost certain mission abort.

Overall, communications between MCC, the CSM, and LM, and communications during lunar EVA were excellent. Most difficulties that were encountered appeared to be procedural problems rather than instrumentation malfunctions. TV transmissions from the CSM and LM and during EVA were amazingly clear, giving people around the world a grandstand view of the most historic event ever witnessed by man.

After a successful liftoff from the lunar surface and rendezvous and docking with the CSM in lunar orbit, the LM crew returned aboard the CSM with the first lunar samples. The TEI burn was executed on the far side of the moon for the return to earth. Again, the trajectory of the CSM was so perfect that midcourse corrections 6 and 7 were not required, and only a slight correction burn was required for number 5. On July 24, 1969 the CM reentered the earth's atmosphere and splashdown occurred at 16:49:00 GMT in view of the prime recovery ship U.S.S. Hornet located approximately 950 nmi southwest of Hawaii.

Experience gained from the previous Apollo missions by the personnel at MSC, GSFC, and the MSFN stations was a real asset in the successful accomplishment of this tremendous feat. The performance of the Apollo 11 crew in the completion of all mission objectives, including deployment of EASEP, was outstanding. Constant communications lock-on with earth by the spacecraft crew and provision for a better view of landing areas during approach should be a goal for future missions.

1-1

"That's one small step for man,..."

Equipment failures, operator errors, and procedural discrepancies and misinterpretations occurred at times during the mission. However, the magnitude of these deficiencies caused no mission impact.

Further comments in this section encompass the entire MSFN. The paragraph numbers listed in the right-hand margins refer to paragraphs within this document where detailed information on particular problems may be found.

Handovers continue to be somewhat of a problem, both from an operational and procedural standpoint. Critical metric tracking data was lost during launch when a station terminated the uplink carrier 30 seconds early. An operator error during an unscheduled handover resulted in a loss of approximately 6 minutes and 30 seconds of active CSM support during the lunar phase. Downlink signal to 7 three-way stations was lost because of this error.   3.1.1.1   3.1.1.3

Numerous systems monitor recorder problems occurred throughout this mission. Eight stations experienced recorded data losses resulting from recorder malfunctions.   3.1.1.5

Five stations encountered 1218 computer problems. While these problems had no significant impact, the program tracking mode of operation was degraded.   3.1.2.1

The USB PA system caused problems at seven stations but had limited mission impact. The problems included: ac overcurrent, arcing beam voltage regulator, loss of beam voltage, local rainstorm, loss of 400-volt three-phase power (short circuit), high ambient temperatures, and motor generator malfunction.

3.1.2.2

Three stations experienced USB paramp malfunctions during the mission. At one station, momentary interruption of tracking was encountered, while at the other two stations, tracking was significantly affected.

3.1.2.3

Timing problems were experienced with the PFS at two stations. One station had a failure of channel 1, while another station encountered oscillations on this channel.

3.1.2.4

Weak and fluctuating signals were reported by all network stations. The causes were attributed to the aspect angle of the spacecraft antennas, the use of the omni antenna at extreme distances, data bit rate changes from low to high, the spacecraft in the PTC mode of operation, and keyhole and terrain masking limitations.

3.1.3.1

During the powered ascent of the LM from the lunar surface, all seven of the supporting stations lost 14 seconds of telemetry data because of the inability to maintain decom lock. The data loss was attributed to the engine exhaust plume impinging on the LM descent stage and lunar surface.

3.1.3.2

Telemetry computers experienced software problems at three stations. There was no significant mission support impact as a result of these problems. Computer faults were encountered at seven of the network stations throughout the mission.

3.1.3.4

Command computer faults occurred at four stations. However, automatic recovery was successfully accomplished with but one exception (MER).

3.1.4.1

During EVA, an echo was evident on the GOSS conference. A modification is being implemented to alleviate this condition. On several occasions throughout the mission, MCC noted distortion on the GOSS conference. The VOGAA is suspected of being a contributing factor and MFED has this problem under investigation. Overall C-band tracking performance more than met mission support requirements. although the average time for establishing track was later than predicted. The main cause for late acquisition appeared to be erroneous IRV's, with side-lobe tracking as a secondary cause.

3.1.5

1.2 NRT PERIOD (TSI 001 TO ISI 001)

1.2.1 GENERAL

The NRT period is used by the ND to determine the status of the MSFN and evaluate it prior to committing it to the support of MCC. The network was placed on premission status by TSI 001 on June 9. At the end of the NRT period on July 6, the MSFN readiness testing had not been completed at all stations for several reasons, including:

a. The lack of complete EASEP equipment configuration documentation being received from the GSFC Requirements Office.

1-3

b. The incomplete installation of required EI's preventing verification of equipment configuration.

c. The reduction of time allocated for stations to configure equipment and perform testing (scheduled time was reduced from 12 days to 8 days).

1.2.2 TRACKING

1.2.2.1 <u>USB</u>. All USB testing was completed with the following exceptions: ACN (IST 02), BDA (ST 10), GDS (IST 02), MAD (ST's 10, 12, and 13), and HTV (IST 02). These stations had all testing complete by July 11.

During NRT, USB encoder corrections to eliminate angular bias were applied to ANG, CRO, GYM, HAW, MIL, and TEX. The corrections were based on analysis of previous Apollo mission tracking data. Subsequent analysis of TETR-B tracks provided data as to the effectiveness of the adjustments.

At the start of the period, MAD's PA No. 2 was inoperative due to a bad voltage regulator. A delay in obtaining a replacement was encountered and the unit remained inoperative until July 8. On this date a new unit was received and installed restoring the amplifier to operation.

The X-axis servo pump at HSK failed on June 26. During repairs it was discovered that the pump port plate on the spare was not compatible with the system. A 5,000-psi plate was installed in the prime pump and the spare contained a 3,500-psi plate. A conference between the manufacturer's engineers and station personnel confirmed the 3,500-psi plate to be the one designed for the X-axis servo pump, and necessary arrangements for installation were made. The unit returned to operational status on July 9. Additionally, arrangements were made to supply spare plates to MAD and GDS.

ANG experienced a failure of the "times 8" multiplier module in exciter No. 2 on July 3. A spare unit was received on July 8 and installed, restoring operation.

A waveguide at BDA began leaking nitrogen at a rate of 600 lbs per day on June 20. Temporary repairs were effected to arrest the leaking of the waveguide and a new waveguide was ordered. The new waveguide did not arrive on station until July 16 and support was provided using the existing unit. Replacement was accomplished post-mission.

The GDSX heat exchanger No. 2 fan was inoperative throughout the NRT period. Support was accomplished with reduced cooling capacity to the PA. Actual impact was dependent on the outside ambient temperature. The unit was repaired on July 12.

GDS encountered a failure of a power cable to the collimation tower in March, 1969, and repairs had not been effected. Support of testing was therefore accomplished with an auxiliary power unit at the tower.

1.2.2.2 <u>Radar</u>. All C-band radars were exercised. VAN experienced a software problem in the CADFISS mode; however, real-time data appeared normal. Two procedural changes had to be implemented to accomplish CADFISS tracking. The first of these was an instruction to bypass IRV time and so permit normal processing. This was necessary because the IRV's used during CADFISS testing contained discrete times instead of all 0's and caused the first two tests to fail. The second change required the use of rate gyros to pass the C-band high-speed test.

Edwin Aldrin Begins His Descent to the Lunar Surface

Tunnel diode amplifier failure, which has degraded support in previous missions, is the only common C-band problem that was not resolved at the end of AS-505. Repeated failures have occurred previously in the MSFN and have reduced C-band tracking capability. It appears that the amplifiers would fail just prior to each mission, be refurbished between missions, only to fail again. This would require reconfiguration of the tracking system and permit limited radar support. The problem appears to have resolved itself, because there were no tunnel diode amplifier failures during NRT, and none were reported during the entire support of AS-506.

1.2.3 TELEMETRY

The installation of EI 3639 caused several stations to incur dump voice distortion problems during testing. The EI was installed to resolve the problem of the decoms being unable to process LM dump telemetry on the CSM FM downlink. This problem was solved, but the degradation of the dump voice resulted. The problem was still under investigation by MFED at the end of NRT and was resolved by ISI 044 during the mission.

During NRT, the Apollo ships were not able to record VHF A/G signal strengths on magnetic tape due to a low input voltage level from the receiver. The problem was resolved by the use of dc amplifiers in series with the data input to the recorder. The Class-19 ships experienced a shortage of tie lines which prevented the recording of Net 1 on the Mincom 22 recorder. Additional tie lines were provided to the ships to resolve the problem. This situation has existed on previous missions, but was only temporarily resolved at the time.

The lack of spare equipment on station to check out the configuration for TV support prevented MAD from completing testing during NRT. Additionally, the MAD TV converter was inoperative due to a defective disc recorder control track head.

The MARS mission support requirements for LM descent and lunar surface activity were received late from MCC and were not tested in a mission configuration until completion of NRT. The MARS testing was completed on July 10. Parkes data flow testing was delayed due to late receipt of mission support requirements from MCC. Late requirements also prevented CRO from testing in a dual Apollo/EASEP telemetry and voice configuration during NRT.

HSK reported an intermittent 15-MHz oscillation of the 642B telemetry computer master clock which made it impossible to master clear the computer when this oscillation occurred. This condition appears when the operating mode is changed (e.g., RUN to OP STEP). When the problem occurs, the master clock can be recycled and remain operative.

On June 26 an OPN was issued to all stations informing them of a problem existing when the DDF-13 is turned off while still connected to the computer. A DDF-13 power supply leakage provides a false external interrupt to the computer. For the mission, the problem was overcome by switching the unit off-line before turning it off.

At the completion of NRT, no major problems existed with biomed, tape recording, or PCM telemetry equipment. An adequate check of the DDF-13 had not been made at GDS, HSK, and MAD due to the incomplete installation of EI 3557 (Change 3) and Errata T-12.

1.2.4 COMMAND

No major command problems were experienced during NRT. The operational command program with Erratas C-1 through C-10, the Erasable Memory Octal Dump (EMOD) program with Errata E-1, and the Off-line Universal Command History (OUCH) program with Erratas 0-1 through 0-17 were used by all stations to provide support.

Both HSK and MAD reported oscillations occurring within the PSK demod section 3 after installation of EI 3046. It was felt the oscillations would be detrimental to the overall UDB performance and Change 4 to the EI was issued to all stations and corrected the problem.

BDA experienced an intermittent command computer external function problem. The 1232 I/O console and HSP printed all output buffer messages and the computer could not be reloaded from either paper or magnetic tape. The problem first occurred on June 19. When EI 3508 was installed, a number of ground wires were connected. One of these wires was not connected properly and was intermittently touching a pin on one of the side connectors. This evidently caused phase 4 of the master clock timing signal to ground out. The improperly connected wire was discovered on June 30, corrected, and the intermittent computer problem did not recur.

CRO reported a command computer fault on June 28 after the computer had been cycling for 12 hours, but diagnostics revealed no problem. GBM experienced a fault on June 19 and shipped a core memory dump of the fault to GSFC for analysis.

The NRT period ended with all remote station command systems operational. The only command system EI outstanding at the time was EI 3557, Change 4.

1-6

1.2.5 AIR-TO-GROUND COMMUNICATIONS

The only problem reported during this period was RED's failure to pass A/G remoting test No. 294. The test was repeated and was passed during the CDDT.

1.3 PRELAUNCH PERIOD (ISI 001 TO LAUNCH)

1.3.1 GENERAL

The MSFN was placed on mission status by ISI 001 on July 7. At this time the ND made the MSFN available to MSC through scheduling procedures for MCC activities. The NC assumed operational control of scheduled facilities in support of MSC requirements.

1.3.2 TRACKING

1.3.2.1 USB. The cooled paramp at ACN required replacement on July 7 because of a low output which degraded the received signal. Replacement was accomplished the same day with no further difficulty encountered.

On June 28, the dc drive motor on the X-Y to AZ-EL convertor at ACN had failed. A spare motor was received on station and installed on July 11 restoring the unit to operation.

CRO experienced a PA failure on July 11 caused by the breakdown of an insulator in the high-voltage power supply. Replacement of the insulator restored the amplifier to operation. On July 13, it again became inoperative, this time due to bad bearings in the motor generator control. Replacement of the bearings resolved the problem.

The timing system at MAD lost synchronization with LORAN-C on July 7. After re-synchronization, the system was operational on July 8.

GWM experienced difficulty in keeping clock "A" and clock "B" synchronized on July 7. Upon instruction from the NST, the fine-tuning circuitry on clock "A" was bypassed. PC cards SQ38 and SP35 were received on July 16 and, after installation on July 17, the timing system returned to operation. The problem recurred postmission and was traced to a factory error; the connection on a PC card had been left unsoldered.

During CADFISS testing on July 11, both the high- and low-speed tracking data from GWM was bad. The USB system was found to have a defective error detect coder which was corrected the same day.

When transmitting, a high VSWR was observed at CYI. The cause of the problem was a defect in the Y-axis rotary joint of the waveguide. The section was removed, overhauled, and replaced on July 15. No further difficulty was encountered with the unit.

On July 11, MAD experienced a problem with the PFS when the servo loop in the frequency combiner would not maintain lock. The servo loop gain was adjusted and the unit was reported "green" on July 12.

GWM experienced a high noise level output of the 70-kHz demod of verification receiver No. 2 on July 14. Two receiver power supply capacitors were replaced which resolved the problem.

1.3.2.2 Radar. No problems were encountered by MSFN radars between the issuance of ISI 001 and launch.

The Plaque Following Its Unveiling

1.3.3 TELEMETRY

When the MSFN was placed on mission status, several problems, tests, mission requirements, and equipment configurations were not resolved. These tasks were accomplished prior to liftoff. Several erratas were transmitted to the MSFN after SI 001 and prior to liftoff. Errata T-12 was generated for PAM DDF-13 telemetry program operation. Errata T-13 was generated to correct an error in the TLM program SLV special processing routine. Any time the IU APS relay thruster data was input on any EMU channel other than 0, the data would not be validated by the program and the output would be static. This errata allows this data to be input and validated on any EMU channel. Errata T-14 was generated to prevent destruction of the quotient register in the telemetry 642B following an abnormal interrupt received from the PCM decom on EMU channel 0.

Among the unresolved tasks, one resulted from a late FSR (No. 26) from MCC which required dual Apollo/EASEP support from CRO, HSK, and MIL. As a result, a test procedure was generated to verify proper configuration requirements.

PAM telemetry interface testing was not completed at GDS, MAD, and HSK until July 8. Installation of EI 3557 (Change 3) and Errata T-12 (DDF-13 word indexing of PAM data in the telemetry operational program) was not incorporated until this time, after which successful PAM EVA testing was accomplished.

Due to requirements for Parkes TV and telemetry support being received late from MCC, the end-to-end data flow equipment interfacing tests between Parkes, HSK,

GSFC, and MCC were not conducted until a few days prior to liftoff. However, the tests were very good and excellent mission support was provided.

A problem area that has existed since the AS-205 Mission is the high failure rate of MSFN TELTRAC acq aid data preamps associated with 0- and 90-degree inputs. Several stations experienced these failures prior to liftoff, but did not encounter any failures during the plus time activity. As a result of this problem, MFED authorized the installation of a modified preamp in the acq aid system at TEX. The modified preamp provides right- and left-hand circular polarization rather than 0- and 90-degree outputs. This configuration results in a more stable preamp due to the deletion of the critical phase-matching network. TEX encountered no preamp problems during the premission phase. The modification has been installed at all MSFN stations with the TELTRAC system.

On launch day, the telemetry CADFISS in the minus count could not be accomplished at GDS until a problem with decom No. 3 was resolved. The telemetry computer would not validate SLV data received from the decom. The PCM error analysis routine in the operational program was used to isolate the problem. An error in the last word of the prime frame synchronization pattern was discovered. The problem was corrected by inserting the correct bit pattern into the PCM memory location and the HBR CADFISS was successfully accomplished.

During the TLM interface test at MIL on July 12, the telemetry computer failed to properly process the LVDC special function set No. 2 data. This data was being processed into the general buffers as opposed to special single-word buffers assigned for this purpose. The problem occurred while playing the AC confidence tape containing IU and S-IVB data into the computer and then disabling the CSM and LM decom inputs. When the decoms were disabled, high-speed error printouts were received for format 13, word 34 (word 34 contains special function set No. 2 data) and format 5, word 169. When the decoms processing LM and CSM data were enabled, word 169 was cleared, but word 34 contents remained the same. As soon as the word 34 error printout started, the RSDP would not process special function set No. 2 data. The problem was temporarily resolved when the M&O faulted the program and accomplished a recovery, then reloaded the program from magnetic tape. The problem recurred during the terminal count, but not during the real-time mission support.

1.3.4 COMMAND

On launch day, both HSK and MIL experienced command history aborts during the terminal count. In both instances, requests for another command history were successful.

During the terminal count, the high-speed command interface was found to be intermittent via the SATCOM satellite due to a noisy SATCOM downlink. VAN was reconfigured to transmit data via SATCOM at 2.4 kbps and to receive data via SATCOM at 1.2 kbps. Command interface was accomplished under this configuration. Additionally, CCATS load control attempted to transmit a command execute to VAN, but transmitted the execute to MAD instead. Subsequent executes were normal.

1.3.5 AIR-TO-GROUND COMMUNICATIONS

MIL A/G receiver switching time during launch was not determined until during the prelaunch period. After much deliberation, it was decided that ISI 095 from AS-505 still applied and the receivers should be switched at launch +1 minute even though a short period of noise would result.

1-9

1.4 LAUNCH PHASE

Liftoff of the Saturn V vehicle occurred following ignition of the S-IC booster stage at 13:32:00 GMT (00:00:00 GET). At the termination of the inboard and outboard engine cutoff, the Saturn V was at an altitude of approximately 36 nmi at 00:02:41 GET. After S-IC/S-II separation, the S-II stage ignition was initiated. The launch escape tower was jettisoned 3 seconds after S-II ignition. The S-II burn continued for 6 minutes and 25 seconds, lifting the Apollo spacecraft and crew to an altitude of approximately 108 nmi. MIL experienced some tracking difficulty because of excessively strong signals and a number   3.19.1 of tracking modes were rapidly attempted to select optimum metric data. VAN did not receive an acquisition message for launch and locked   3.23.2.1 on to a carrier sideband so that two-way data was degraded until re-acquisition was achieved. Ignition of the S-IVB stage was initiated 3 seconds after S-II separation. At 00:11:39 GET, the S-IVB engine cut off and 10 seconds later EOI was achieved with a parking orbit altitude of 103.3 nmi. BDA then successfully transmitted the safing command to   3.6.4.2 the S-IVB at 00:12:04 GET. GBM discovered postpass that the output of the data modem had not been recorded because of a patching error.   3.11.1

1.5 EARTH ORBIT PHASE

The Apollo 11 spacecraft remained in an earth parking orbit for approximately 2 hours and 32 minutes. During EO revolution 1, GDS   3.10.2.1 experienced difficulty with a TV transmission from the spacecraft because of a defective patch cord in the SDDS. GDSX was unable to   3.10.2.2 provide TV support for this revolution because of terrain masking. TAN C-band tracking was erroneous during this EO revolution be-   3.21.2.2 cause of improper adjustment to find and verify threshold levels. During the second EO revolution, a failure of a LDN receive cir-   3.21.5 cuit caused a loss of MCC uplink voice through TAN.

1.6 TRANSLUNAR COAST PHASE, INCLUDING TLI

During EO revolution 2, as the spacecraft was over the Pacific near the Gilbert Islands and approaching Hawaii (02:44:15 GET), a second ignition of the S-IVB engine was initiated. This burn duration was for approximately 5 minutes and 48 seconds, and increased the velocity of the S-IVB enabling it to escape earth orbit and be injected into a translunar trajectory. MER was unsuccessful in providing required TLI tracking support because of a malfunctioning command computer. The handover   3.18.2.1 from RED to MER was delayed by this problem. RED provided good metric tracking data and monitored S-IVB ignition and the early portion of the TLI burn as required. MCC initiated a successful contingency handover from MER to HAW at 02:49:04 GET, but some data was lost by   3.18.2.1 MER prior to this time (TLI ignition at 02:44:16 GET). C-band radar support was limited during TLI because of MER's failure to comply with   3.18.2.2 ISI 053, resulting in late acquisition. Acquisition was finally accomplished by slaving to the USB antenna, but the TLI burn was approaching cutoff at this time.

During the first view period of TLC, several uplinks through GDS   3.10.5.1 were not completed. It was determined that the problem did not exist at GDS but somewhere between GSFC and MCC. Downlink voice remoted to MCC by GDS was distorted at approximately   3.10.2.1

Erecting the Solar Wind Component Experiment

02:55:00 GET. This difficulty was caused by a defective input cable to a voice demod. Additionally, a substandard amplifier was found in another voice demod.

GDS failed to remote downlink voice to MCC at 03:00:00 GET for the first 10 minutes of the view period. A misinterpretation of NOD instructions was responsible for the failure of this downlink reaching MCC. At 04:00:00 GET, remoted voice was distorted because of a defective cable to the demod. Handover to MAD was required.

3.10.5.1

Approximately 27 minutes after TLI at 03:15:01 GET, the S-IVB and the CSM separated in preparation for the extraction of the LM from the SLA. After nominal transposition, docking, and extraction, a CSM/LM evasive maneuver was initiated at 04:39:45 GET to ensure against any possible recontact with the S-IVB vehicle. The Apollo crew was able to visually monitor the S-IVB during this maneuver.

At approximately 2 hours after TLI, the S-IVB was commanded by MCC to assume a local horizontal attitude for final fuel expulsion, a maneuver which further reduced the probability of recontact with the CSM/LM and ensured that the S-IVB would not impact on the earth or moon. The fuel dump placed the S-IVB into a "sling-shot" trajectory so that it passed behind the trailing edge of the moon. Taking advantage of the lunar gravitational field, the acceleration of the S-IVB increased, placing it into a heliocentic orbit.

BDA C-band radar tracked the S-IVB/IU beacon for 8.5 hours to a range of more than 52,000 nmi. Difficulty was experienced with VHF support during TLC because of RFI received when the antenna tracked the spacecraft into the path of the sun.

3.6.2.2

3.6.3.2

During HGA handover to MAD at 04:02:00 GET, the exciter operator, in error, used omni antenna handover procedures. Because of this error, MAD was required to initiate uplink sweep to acquire. Later in this view period between 04:40:06 and 06:23:00 GET, the low- and high-speed data from the TDP was flagged "bad." Because of an incorrectly positioned switch, the data was invalidated.

3.10.2.1

At 20:50:00 GET, tracking was momentarily interrupted at HSK when the antenna slewed off in the X-axis while autotracking. This difficulty was caused by a corroded connector on the X-axis channel cable.

3.15.2.1

Several command MTU difficulties were experienced at MAD during TLC (05:13:00, 24:02:04, and 28:20:00 GET); RTC histories were lost on each occasion. During the first problem occurrence, the command information was extracted from the HSP and sent to MCC by teletype. Replacement of flip-flops and amplifier cards resolved an MTU difficulty later in the mission.

3.17.4

At CYI the radio telescope associated with the SPAN system became inoperative at 25:13:00 GET. Repairs were completed and the radio telescope was operational at 33:58:00 GET. The radio telescope at CRO failed because of a broken shear pin in the antenna at 33:52:00 GET. Replacement of the shear pin resolved the problem and operation was restored at 34:51:00 GET.

3.2.2

3.2.2

The supply control system of PA No. 2 at HSKX was severely damaged at 42:52:00 GET when an electrical fire resulted from a short circuit in the 400-volt, three-phase power supply. Uplink capability was lost, and PA No. 1 was brought into service.

3.15.2.2

MADX Doppler data was biased during TLC (45:46:37 GET) by an incorrectly selected frequency on the frequency synthesizer. The error was detected and corrected after an elapsed time of approximately 1 hour.

3.17.2.2

On two occasions MADX had problems with high ambient temperatures in the hydro-mechanical building (47:10:00 and 48:18:10 GET). Uplink capability was interrupted both times resulting in a handover to the prime station.

3.17.2.2

At 46:58:00 GET, valid EKG data from MAD was not received at MCC because of a patching error. The station was informed of the error and reconfiguration corrected the problem.

3.17.3.1

Ten minutes of real-time telemetry data was lost at ACN (70:16:00 GET) when the telemetry computer faulted. A reversed computer configuration was necessary to restore support capability.

3.5.3.3

The CSM/LM continued on toward the moon and the accuracy of the trajectory was such that the first, third, and fourth midcourse cor-

1-12

rection burns were not required. A slight correction was required at approximately 26 hours and 45 minutes into the flight (second midcourse correction). During TLC, 3 hours and 37 minutes of color TV was received from the CSM and remoted to MCC through GDS, MAD, and the MARS station.

1.7 LUNAR PHASE

On July 19 at 75:49:50 GET, a 6-minute SPS burn (LOI-1) placed the CSM/LM into an elliptical orbit approximately 167 by 60 nmi above the lunar surface. Four hours and 22 minutes later, after 2 orbits of the CSM/LM, a second SPS burn (LOI-2) circularized the orbit to approximately 65 by 52 nmi with the perilune placed at approximately 70 degrees west longitude. This orbit was so designed that variations in the lunar gravity would circularize the orbit to 60 by 60 nmi at the time of LM ascent some 44 hours later. At 95:20:00 GET, the CDR and LMP entered the LM to check out the vehicle systems and make the necessary preparations to undock from the CSM.

During a contingency handover to MAD (76:27:00 GET), 37 minutes after LOI-1, ACN received six ground rejects for uplinked commands. The commands were attempted during the time that the carrier was not being modulated.   3.5.4

At 84:44:00 GET, a power supply failure caused the loss of receiver No. 1 at GDS. The exciter operator initiated two-way reacquisition while receiver No. 2 was in lock, sweeping the exciter before shorting the synthesizer loop. This error caused the spacecraft HGA to slew off. Reacquisition procedures were initiated but caused delayed AOS.   3.10.2.1

After completion of the systems checkout, and approximately 5 hours after the crew entered the LM, undocking occurred at 100:14:00 GET during LO 13. After a brief 26 minutes of station keeping, an RCS burn of the CSM separated the two vehicles.

A procedural problem was experienced by ANG at 101:13:00 GET after the CSM separation burn and prior to the LM DOI burn. The ComTech was receiving downlink from the spacecraft but no uplink on Net 1. GSFC had removed ANG from Net 1 because of false rings on the circuit.   3.4.1

One half revolution after CSM/LM separation and one half revolution prior to powered descent to the lunar surface, the DOI burn was initiated at 101:36:14 GET. This maneuver placed the LM in a 60-nmi by 50,000-foot orbit with perilune east of the landing site. The DOI consisted of one RCS ullage burn of 7.5 seconds and two DPS retrograde burns of 15 seconds (12-percent throttle) and 14 seconds (40-percent throttle). PDI occurred at 102:33:04 GET when the LM was in the perilune of the descent transfer orbit. The descent consisted of three phases: braking, approach, and landing. Maximum thrust from the DPS was required during the braking phase, which was designed for the efficient reduction of the orbital velocity; however, the DPS was throttled during the final 2 minutes of this phase for guidance control. At approximately 7,000 feet above the lunar surface and with the LM attitude such that the crew could visually monitor through the forward windows, the approach phase was initiated. During the final computer-controlled

1-13

"...The heavens have become part of man's world..."

landing phase, the landing point was observed to be in a large crater filled with boulders up to 10 feet in diameter. With only seconds until touchdown, the crew elected to fly manually to a smooth landing point beyond the crater and approximately 4 miles down range from the planned landing site. After probe contact with the lunar surface, the descent engine cut off and landing occurred at 20:17:45 GMT on July 20, 1969, 102 hours, 45 minutes, and 45 seconds after liftoff from Cape Kennedy. The landing point coordinates on the lunar surface were approximately 0.69 degrees north and 23.46 degrees east. The first words from the moon were those of Commander Armstrong: "Tranquility base here, the Eagle has landed." From the command ship Columbia, orbiting 60 miles overhead, Michael Collins added "Fantastic." The CSM participation in the mission, while not as colorful, was equally important because this craft's objective was to return man safely to earth.

After LM touchdown, a hydraulic failure occurred at GDSX at 103:18:40 GET. Brakes were applied while the antenna was still on track and so prevented data loss. Repair was accomplished and autotrack resumed.

3.10.2.2

The crew's observations from within the LM indicated that the tilt of their vehicle was 4.5 degrees from vertical and it was yawed left approximately 13 degrees. The landing had been a smooth one with a soft touchdown. Numerous boulders up to 2 feet in diameter were in the field of view from their spacecraft window. The color of the lunar surface varied from light to dark gray. They observed that some boulders had been fractured by the exhaust from the descent engine. These boulders appeared to be coated light gray and the fractures were much darker. There was a hill in view at a distance of between 0.5 and 1 mile in front of the LM. The crew found that they could adapt well to the one-sixth lunar gravity within the spacecraft. The astronauts were so elated with the mission progress through the lunar landing that, after completion of scheduled objectives, a request was submitted to MCC to reschedule the EVA to take place before their sleep period. MCC granted this request and EVA was scheduled to take place approximately 4.5 hours earlier than planned. The sleep period was rescheduled to take place after completion of the EVA.

On July 21 at 02:39:35 GMT (109:07:35 GET), the LM hatch was opened in preparation for the CDR to initiate lunar EVA. Assisted by the LMP, the CDR egressed the LM and was on the porch at 02:48:19 GMT (109:16:19 GET). He slowly descended the ladder attached to the craft's right front landing leg and at 02:56:20 GMT (109:24:20 GET), the foot of man, that of Neil A. Armstrong, for the first time in history, touched the surface of another planet. MCC received the first words from man standing on the lunar surface, "That's one small step for man, one giant leap for mankind." The CDR made a brief inspection of the LM exterior and observed that footpad penetration was only 1 to 2 inches into the surface and strut collapse was minimal, which indicated that the landing was a soft one and that the surface was firm enough to adequately support the spacecraft. He further reported that he sank only about 1/8-inch into a powdery surface material which adhered readily to his lunar boots. There was no crater created from the effects of the descent engine blast, and the engine bell clearance from the surface was approximately 1 foot. The CDR also reported that darkness in the shadows made it difficult at times for him to see his footing. After a brief period of picture taking, the CDR collected the contingency sample near the LM ladder. He discovered that at a depth of 6 inches into the lunar surface, the material was very hard and cohesive.

Eleven minutes later at 109:35:20 GET, LMP Edwin Aldrin descended the ladder to the lunar surface. The CDR photographed his egress and descent. The CDR and LMP unveiled and read a plaque which bore the signatures of President Nixon and the Apollo 11 crew. After the completion of this event, the CDR repositioned the TV camera for a better view of the remaining EVA. The solar wind component experiment was deployed by the LMP near the LM. A 3- by 5-foot American flag was erected on an 8-foot aluminum mast by the LM crew. Upon the completion of this event, a telephone call from President Nixon, 250,000 miles away, was relayed through MCC. The President said, "I simply can't tell you how proud we all are; because of what you have done, the Heavens have become part of man's world. ...Our prayers are that you return safely to earth." The crew then continued the collection of bulk samples. Further inspection of the LM was made and it was found that no damage had occurred during the landing. The passive seismic experiment package and the laser ranging retro reflector were deployed to the south of the LM. Performance of the package was excellent; the crew's footsteps on the lunar surface and later in the LM were detected by the instrumentation.

More lunar samples were collected including two core samples. The LMP had to exert considerable force to drive the core tubes 8 to 9 inches into the surface.

Four hours and 55 minutes of black and white TV provided real-time coverage of the crew's activity on the surface of the moon and also provided time correlation for telemetry and voice comments. The Parkes station in Australia provided the bulk of the coverage and remoting with an early assistance from MARS and HSK. Most important however, TV provided live documentation of the most historic event and achievement in the history of man. Other lunar surface photography was provided by sequence, data acquisition, and special close-up cameras.

After 2 hours and 14 minutes, the EVA was terminated with the transfer of samples and film to the LM. The crew's ingress was followed by the jettisoning of equipment no longer required. These events were accomplished nominally according to the mission plans. The crew's scheduled post-EVA rest period was expanded to include that time postponed from the previous scheduled rest period which was not used because of the early lunar EVA.

A handover to HSK was scheduled for GDSX during the lunar stay. At the time, the LM was downlinking FM on the HGA. With no procedure documented for this handover, a procedure was developed in real time that proved successful.   3.10.2.2

With the termination of the extended rest period, the CDR and LMP activated all required systems and made the necessary tests preparatory for liftoff from the lunar surface. The CMP in the CSM orbiting overhead maneuvered his craft to support the LM liftoff. At 17:54:00 GMT (124:22:00 GET) on July 21, the LM ascent engine ignition was initiated and a successful liftoff was accomplished without difficulty. From liftoff through a successful rendezvous and docking with the CSM, the performance of the astronauts was outstanding. The rendezvous radar tracked the CSM successfully and all maneuvers of both the CSM and LM were nominal. CSM/LM docking occurred at 128:03:00 GET, 3 hours and 41 minutes after the LM lifted off from "Tranquility Base."

Several seconds of critical LM data was lost by MAD at 126:42:00 GET   3.17.3.1
because of the incorrect selection of a data demod bandwidth. MCC
selected ACN data until the problem was resolved. Just prior to LM
and CSM docking at 127:57:40 GET, the downlink signal acquisition   3.10.2.1
was delayed for approximately 6 minutes and 28 seconds by GDS. The
exciter operator did not use the contingency procedure according to
the NOD. Narrow-loop bandwidth was selected prior to turning off the
exciter sweep bias and thus caused loss of signal. It was necessary to
reacquire using wide bandwidth.

Post-docking activities included the cleaning of equipment, sample containers, and the CDR's and LMP's garments and boots. Precautions were taken to ensure that any contamination resulting from lunar contact would be minimal. After preparing the LM ascent stage for jettisoning, the LM crew reentered the CSM with their lunar samples. The ascent stage was jettisoned at 130:10:00 GET, 2 hours and 7 minutes after CSM/LM active docking. The LM ascent stage was tracked until 138:04:00 GET, at which time CSM tracking requirements became prime. Results of tracking the LM indicated an orbit decay of 2 kilometers per orbit and a lifetime expectancy of approximately 41 orbits, or 4 days after tracking termination.

Following the crew's eat period, preparation was initiated for the TEI burn required to place the CSM into a trajectory to escape lunar orbit and bring the Apollo 11 crew home.

1-16

EASEP Deployed on the Moon

1.8 TRANSEARTH COAST PHASE, INCLUDING TEI

With all systems tests successfully completed, ignition of the SPS engine was initiated at 135:24:34 GET with a burn time of approximately 2 minutes and 30 seconds. The CSM was then in TEC and approximately 59 hours and 40 minutes from earth entry interface (400,000 feet above the earth's surface). A slight midcourse correction (MCC-5) was required at TEI + 15 hours (150:29:56 GET), but the remaining corrections (MCC-6 and -7) were not needed because of the accuracy of the CSM trajectory.

Three color TV transmissions totalling 31 minutes were beamed earthward during TEC. The first telecast, which lasted 18 minutes, was remoted to MCC through GDSX at 155:36:00 GET. A second brief color telecast of 3 minutes was received at MAD and GDSX and was remoted to MCC via GDSX at 177:10:00 GET. The final 12-minute color telecast was remoted through GDS and GDSX at 177:31:00 GET.

RTC was unable to command an omni antenna switch through HSK at 152:50:00 GET. A 10-minute loss of voice resulted before successfully completing the command switch. At approximately 168:15:01 GET, the exciter bias at HSK was delayed in wide bandwidth by operator error while the receiver was in narrow-loop bandwidth. The error caused a momentary loss of signal at HGA handover.

3.15.5

3.15.1

1.9 REENTRY AND SPLASHDOWN

The Apollo 11 CSM, only a few hours away from splashdown in the Pacific, was on a perfect course to complete successfully the most history-making voyage ever attempted by man.

1-17

At 194:49:20 GET, separation of the CM/SM occurred. The SM had fulfilled its prime objective of taking the Apollo 11 crew into a lunar orbit and returning them safely again to the vicinity of earth. Within the span of only a few minutes after separation, the SM burned as it plunged into the earth's atmosphere.

RED had a secondary mission requirement for C-band radar to track the SM during reentry. Because of incorrect acquisition information (CM would precede the SM; radar to track trailing target), RED tracked the trailing target according to instructions, but this target was the CM.

3.20.2.2

Fifteen minutes after separation, the cone-shaped CM entered the earth's atmosphere at 400,000 feet above the Pacific Ocean. At 195:13:00 GET, the drogue parachutes were deployed to slow the CM. One minute later the main parachutes were deployed at approximately 10,000 feet to lower the CM to a splahsdown point within view of the prime recovery ship, U. S. S. Hornet. Splashdown occurred at 16:49 GMT (195:18:28 GET) on July 24, about 950 miles southwest of Hawaii. The landing point had been shifted 250 miles to the east because thunderstorms and choppy seas were forecast for the prediced landing area. The crew made this final correction by shifting the angle of the spacecraft as it entered the earth's atmosphere. In this decade men had walked on another planet and had returned safely to earth, bringing with them material proof of their accomplishment.

SECTION 2. NETWORK PERFORMANCE AND STATISTICAL ANALYSIS

2.1 GENERAL

During AS-506, a total of 5,727 individual data items were received by DSS at GSFC from the MSFN stations, ships, and aircraft. DSS, the collection and distribution center for the network, processed and distributed a total of 6,259 data items. Of these, 706 pieces of data shipped were originals and 5,553 pieces of data shipped were copies reproduced from originals. A listing of data items received by DSS from the MSFN is shown in table 2-1.

Few data handling problems occurred during the mission. Last-minute changes to data handling instructions were minimized, largely due to a premission conference between DSS and MSFN stations via SCAMA. As in past missions, the primary problem of data shipping resulted from shipments missing scheduled commercial airline flights and arriving late at DSS.

During AS-506, the MSFN performance was generally similar to that of previous lunar missions. The USB network, overall, performed essentially the same with the exception that during TEC, performance was slightly below what was expected. The C-band network performance was essentially the same as during AS-503 and better than during AS-505.

2.1.1 EXPLANATION OF STATISTICS

The statistics presented in this section are for the purpose of providing a general description of the performance of the network. In some cases, factors other than the performance alone may affect the statistics; thus, careful reading of the explanation of these statistics is recommended in order to avoid misinterpretation.

2.1.2 DEFINITION OF STATISTICAL TERMS

2.1.2.1 <u>Mean</u>. As used in this report, the mean or average percent for a given category will refer to the arithmetic average of the percents for each view period in that category.

2.1.2.2 <u>Standard Deviation</u>. The standard deviation (SD) is derived by the formula

$$\sigma = \sqrt{\frac{\sum (X_i - \bar{X})^2}{n}}$$

where X_i is the percent for each view period, $\bar{X}$ is the mean for the distribution being considered, and n is the number of data points in the distribution. The SD indicates how closely the points in the distribution cluster about the mean and is used exclusively for comparison between distributions. For example, given distributions A and B with identical means, but with A having a larger SD than B, we know that A has more higher and lower points than B. It must be noted that since the difference between each point and the mean is squared, more weight is given to extremely large differences than to small differences.

2.2 TRACKING

2.2.1 USB

Overall, MSFN S-band tracking system performance was acceptable. The RTCC and GRTS computers received sufficient metric tracking information to provide real-time computation of trajectory parameters during all phases of the mission. Of the data received, most was accepted and processed to establish the spacecraft trajectories and to determine the orbits.

Table 2-1. Mission Data Received--by Station

| Misc | PLIM's | Paper Tapes | Strip-charts | Magnetic Tapes | | | | | | Total |
| | | | | Analog | | Digital | | | | |
				WB	Voice	Log	Raw	Radar	CMD Hist	
2	211	-	41	89	15	-	-	-	1	359
26	144	-	33	13	16	-	-	-	-	232
42	5	-	-	18	6	-	-	-	-	71
28	197	44	41	16	24	-	-	4	1	355
2	6	-	5	-	1	-	-	20	-	34
16	173	-	38	30	18	-	-	2	1	278
2	115	-	22	20	22	-	-	-	1	182
2	21	-	5	16	2	-	-	-	-	46
145	219	-	47	16	18	-	-	-	2	447
3	65	-	41	127	-	-	-	-	-	241
27	177	-	40	91	22	-	-	-	1	358
8	192	-	35	17	18	-	-	-	-	270
13	262	42	47	72	18	-	-	1	1	456
108	199	-	65	114	18	-	-	-	1	505
7	88	-	40	130	-	-	-	-	-	265
4	6	-	12	2	-	1	-	-	-	25
35	179	-	40	100	16	-	-	-	2	372
9	67	-	41	101	-	-	-	-	-	218
10	14	3	8	6	1	2	1	-	-	45
85	232	9	153	131	67	-	-	-	4	681
-	-	-	-	19	-	-	-	-	-	19
-	-	-	-	4	-	-	-	-	-	4
2	8	-	2	4	1	-	-	1	-	18
6	127	-	36	12	9	-	-	-	-	190
9	24	4	9	4	1	2	2	-	1	56
596	2731	102	801	1152	293	5	3	28	16	5727

2.2.1.2 Deviation. The USB tracking network was analyzed statistically using three performance indicators. An explanation of the derivation of these indicators and of factors which should be considered when attempting to interpret them is as follows:

a. Acquisition Deviation Between Predicted and Actual AOS. This value was derived by subtracting the AOS time listed on the USB PLIM from the predicted AOS time listed on the GCC PLIM. Variations in acquisition can be attributed to several causes, the most obvious being lack of operator proficiency and equipment capabilities. However, other significant variables exist. Early acquisition may result from favorable propagation conditions due to ionospheric anomalies, or apparent early acquisition may occur because of an acquisition message error in the predicted time. Late acquisition can be caused by several factors beyond the control of the station. Horizon masking, because of terrain irregularities at certain azimuth approach angles, is inherent in the location of some stations. Variations in atmospheric conditions affect propagation conditions. Particularly at low antenna angles, acquisition condition can have either an adverse or a favorable effect. An error in the computer-generated acquisition message or a delay in receiving the message will delay radar acquisition by extending the search required; at worst, the search can preclude acquisition during that pass. The 0-degree elevation angle intercept point, used by the computer to derive acquisition time predictions, is not normally achieved because of the foregoing limitations and because of signal fluctuations due to multipath reception at low elevation angles within one-half beamwidth of the horizon, which prevent continuous lock and valid signal indications. Because of these factors, acquisition deviations should not be considered a measure of station performance exclusively.

b. Percentage of Actual Track to Computer-predicted View Time. This value was derived by subtracting the AOS time from the LOS time listed on each USB PLIM, dividing it by the difference between the computer-predicted AOS and LOS times listed on the corresponding GCC PLIM, and multiplying by 100. The computer predictions for LOS are developed on the same basis as those for AOS (0-degree elevation angle) and are therefore influenced by the same factors discussed above for AOS predictions.

Since differences in length of view period during different phases affect these percentages, actual-to-predicted statistics from different phases should not be compared directly. For example, during launch and EO, a 90-percent actual-to-predicted time would mean that the actual time tracked differed from the predicted time by less than a minute. During the orbital portion of the lunar phase, it would mean that the actual track was about 7 minutes less than predicted, and during one of the coast phases this 90 percent would mean that the actual track could be 1 or 2 hours less than predicted. Thus a 90-percent actual-to-predicted time for an earth orbit would be better than the same percentage for a lunar orbit which would, in turn, be better than 90 percent for either of the coast phases.

The SD is also affected by the length of view period. For example, during a 10-minute view period, a 30-second change in actual time would produce a change of 5 percent in the percentage of actual-to-predicted time, whereas the same change during a 10-hour view period would produce a change of less than 1 percent. Thus large fluctuations in the actual or predicted times during a coast phase will not result in as large fluctuations in percentages as would these same fluctuations during earth or lunar orbit. Therefore, as in the case of the averages, a given SD represents more consistency in performance during EO than it would during the orbital portion of the lunar phase, and better performance than during a coast phase.

c. Percent of Valid Track to Total Track. These percentages were derived by adding two- and three-way track lines listed on each USB PLIM, multiplying by six lines per second, dividing by the difference between AOS and LOS listed on the same

PLIM, and multiplying by 100. This percentage is probably the best single measure of the tracking performance.

The valid-to-total track percentages are affected by the variation in length of view period in basically the same way as the actual-to-predicted percentages.

The time that it takes to achieve lock after AOS is a smaller percentage of the total track time during a coast phase than during an earth or lunar orbit. A few dropouts do not have so great an effect on the percentages during the coast phases as during orbit, and the percentages during the coast phases are not so sensitive to fluctuations in actual times as during orbit. Therefore, direct comparison among phases by use of these statistics is also inadvisable except between the two coast phases.

For calculations of mean and SD for the different indicators, cutoff points were chosen in order to bring the data points more in line with practicality and to exclude obviously erroneous data. Cutoff points were as follows:

(1) Deviations of AOS from predicted that were greater than 400 seconds; 53 data points were excluded, 34 of them from TLC and TEC.

(2) Percentages of valid track to actual track that were less than 10 percent; nine points were excluded.

(3) Percentages of actual track to predicted view that were greater than 150 percent; five points were excluded.

Table 2-2 shows the number of points excluded from each station for each phase. Table 2-3 summarizes the overall network tracking performance by phase for AS-506. Except for the coast phase statistics, the statistics for a given phase should not be directly compared to those of any other phase.

d. <u>Launch and EO.</u> During launch and EO, the network averaged earlier than predicted acquisition and tracked longer than predicted. While the SD's indicate this was not consistently the case, the two averages would imply that either the predictions were slightly conservative or ionospheric conditions were such that several stations were able to acquire signal while the spacecraft was still below the horizon.

e. <u>TLC.</u> Although acquisition averaged late during TLC and deviations fluctuated greatly, the network on the average tracked for most of the predicted view time and did so fairly consistantly. The fact that the late acquisition and extreme fluctuations in deviation between actual and predicted AOS did not result in a lower percentage of actual track to predicted view or a large SD for these percentages can be explained by the fact that deviations of less than a minute during a coast phase do not change the percentage of actual track to predicted view by even 1 percentage point.

f. <u>Lunar Phase.</u> Acquisition during the lunar phase was also later than predicted, but deviations were somewhat more consistant. There were consierably more data points available for calculating the lunar phase deviations than for either of the coast phases. Generally the more points used for a given data set, the smaller the SD.

The mean percentage of actual track to predicted view time for the lunar phase appears only slightly better than that for TLC and TEC. However, due to the effect

2-4

of the length of the view period on the statistics, it actually represents significantly superior tracking performance. For this same reason, the SD for the lunar phase, which is somewhat larger than that for TLC, probably represents more consistency in tracking performance. This same principle holds true for valid to total track time. The statistics, which are essentially the same, actually represent tracking performance for the lunar phase significantly superior to TLC.

Table 2-2. Points Deleted

Station	Deviation				Act/Pred				Val/Total				Total
	EO	TLC	LUNAR	TEC	EO	TLC	LUNAR	TEC	EO	TLC	LUNAR	TEC	
ANG		2	1	2									5
ACN			1	1									2
BDA			1										1
CYI		1											1
CRO			1										1
GDS		2	4	2	2		1				2		13
GDSX			2	2									4
GBM													
GWM				1			1						2
GYM			3				1						4
HAW		1									1	1	3
HSK		2	2	1							1		6
HSKX		2	2	2			2						8
HTV				1									1
MAD		4	1	2									7
MADX		3		2							1		6
MER									1				1
MIL			1	1							1		3
RED													
TEX											1		1
VAN													
Total		17	19	17	2		5		1		7	1	69

2-5

Table 2-3. Statistical Summary of the Network USB Tracking
Performance During AS-506

	Acq Deviation Between Predicted and Actual AOS (Seconds)		Percentage of Actual Track to Predicted View		Percentage of Three-way Plus Two-way Track to Total Track	
	Mean*	SD	Mean	SD	Mean	SD
Launch and EO	+2.4	36.2	103.7	23.2	79.2	14.3
TLC	−49.7	149.1	95.6	8.7	90.2	11.3
Lunar Phase	−32.2	85.0	97.4	12.1	90.9	14.8
TEC and Reentry	+61.2	137.8	94.9	17.8	85.2	17.5

*The minus signs indicate acquisition later than predicted.

g. TEC. During TEC the network averaged early rather than late acquisition as during TLC, but with the same extreme variation as during TLC. The mean for actual track to predicted view shows that, over all, the network performance was essentially the same during TEC as during TLC, with the SD for TEC showing less consistency of performance than during TLC. With respect to valid track to total track, the network performance during TEC was definitely inferior to that during TLC.

h. Graphic and Tabular Representation of Actual S-band Track to Predicted View.

(1) Percentage of Actual Track to Predicted View. Figure 2-1 gives in finer detail the data points used in the calculation of percentage of actual track to predicted view, again divided into four phases:

(a) Launch and EO (Figure 2-1a). During launch and EO, 78.9 percent of the reported tracks exceeded the predicted track times and 68.4 percent exceeded the phase average.

(b) TLC (Figure 2-1b). During TLC, 29.1 percent of the reported tracks exceeded the predicted track times and 70.9 percent exceeded the phase average.

(c) Lunar Phase (Figure 2-1c). During the lunar phase, 56.9 percent of the reported tracks exceeded the predicted track times and 78.2 percent exceeded the phase average.

(d) TEC and Reentry (Figure 2-1d). During TEC, 42.8 percent of the reported tracks exceeded the predicted track times and 71.4 percent exceeded the phase average.

The fact that during all phases well over 50 percent of the reports are above the phase average indicates that a few low percentages are inordinately affecting the averages.

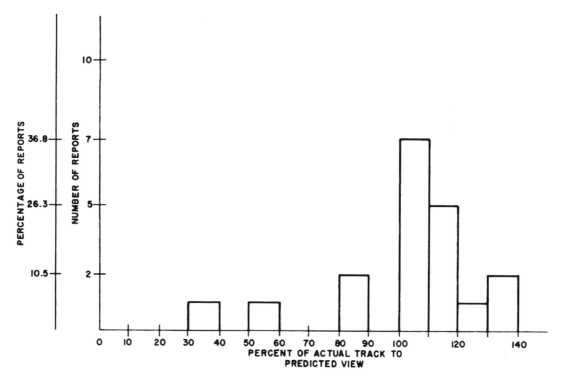

a. Launch and EO

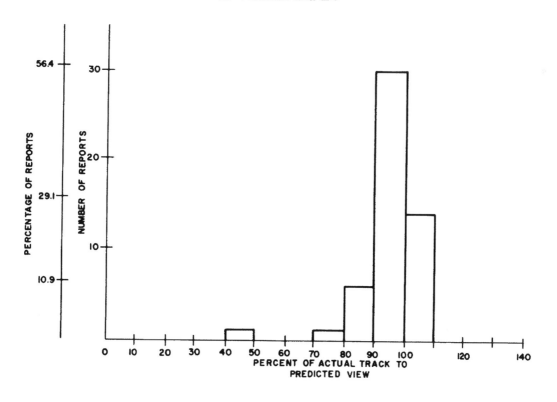

b. TLC

Figure 2-1. Actual S-band Track to Predicted View

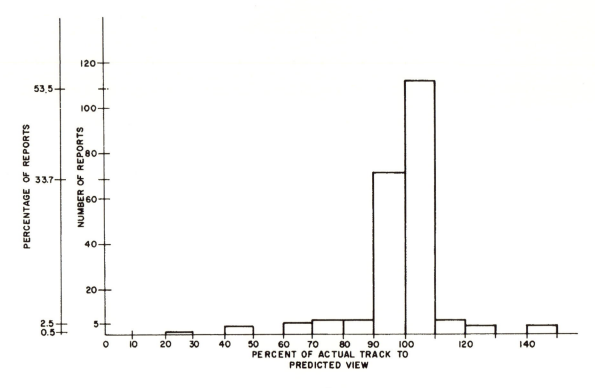

c. Lunar Phase

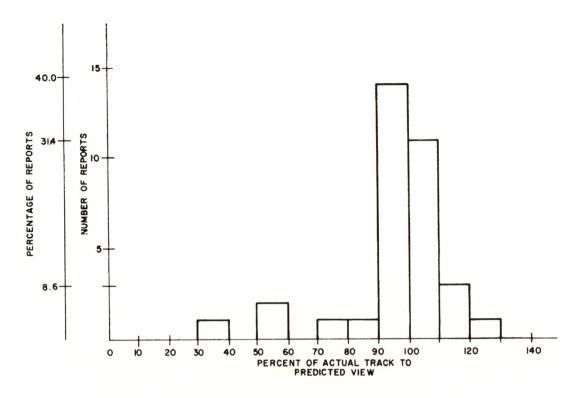

d. TEC and Reentry

Figure 2-1. Actual S-band Track to Predicted View (cont)

(2) <u>Percentage of Valid Track to Total Track.</u> Figure 2-2 is a more detailed presentation of the data used in calculating the percentage of valid track to total track for the mission as a whole. The mean of 89.2 percent for the network was surpassed 69.0 percent of the time in which the vehicles were being tracked. Once again, the fact that well over 50 percent of the reports exceeded the average shows that a small number of low percentages has had an excessive effect on the average.

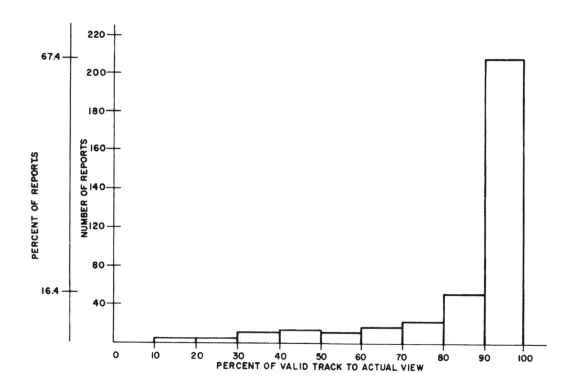

Figure 2-2. Percentage of Valid S-band Tracking to Total Track

(3) <u>Individual Station Performance</u>. Table 2-4 details the tracking performance for the mission from still another perspective. It presents the individual station performances during each phase. The following observations can be made of the data contained in this table:

(a) Thirteen of the 21 stations (this includes wing stations separately) tracked at some time during launch and EO; of these stations, nine averaged acquisition times earlier than predicted, and four averaged later. Nine stations averaged longer track times than predicted, and four averaged shorter times than predicted. The column listing the average percentage of valid track reveals that six stations were above the network average for these phases. One station had no valid track.

(b) Sixteen stations tracked during TLC. Of these, two averaged early acquisition times, 13 averaged late, and one had acquisition times which deviated from the predictions by more than 400 seconds. Four stations tracked for a longer time period than predicted, and all but two averaged actual track of over 90 percent of predicted view times. Twelve stations averaged valid track for over 90 percent of their actual track.

2-9

(c) Sixteen stations tracked during the lunar phase. Of these, two averaged early acquisition times and 14 averaged late. Three stations tracked longer than predicted and 10 had valid track for more than 90 percent of their actual track time.

(d) Sixteen stations tracked during TEC and reentry. Of these, eight averaged earlier acquisition times than predicted, three later, and five had acquisition times which deviated from the predictions by more than 400 seconds. Six stations tracked for a longer time period than predicted. Eight stations obtained valid track for more than 90 percent of their actual track, and four had valid track for less than 65 percent of their track time. Two of those having extremely low percentages of valid to actual track were ships.

Table 2-4. Statistical Summary of USB Tracking Performance
by Station During AS-506

Launch and EO				Lunar Phase			
Station	PAOS-AOS (Sec)*	% of Act/Pred Track	% of Valid/ Total Track	Station	PAOS-AOS (Sec)*	% of Act/Pred Track	% of Valid/ Total Track
BDA	+6.5	104.1	82.9	ACN	−26.6	93.2	88.4
CRO	+67.5	123.8	76.6	ANG	−26.9	99.2	85.7
CYI	+1.5	115.6	74.5	BDA	−38.7	99.5	92.3
GBM	+12.0	95.0	85.1	CRO	+18.9	104.8	91.5
GDS	−76.0	56.0	76.4	CYI	−22.0	99.5	97.9
GDSX	−23.0	38.9	33.0	GDS	−35.9	94.9	89.2
GYM	+3.0	114.1	86.3	GDSX	+5.8	92.5	99.1
HSK	−58.0	81.5	78.8	GWM	−47.9	100.5	90.7
MER	+35.0	118.8	0	GYM	−30.0	99.3	92.5
MIL	−4.5	109.1	80.5	HAW	−17.9	99.8	85.6
RED	+4.0	108.9	74.3	HSK	−87.0	87.3	86.9
TEX	−54.0	117.3	81.3	HSKX	−54.2	89.8	86.2
VAN	+15.0	108.9	74.3	MAD	−37.3	94.7	90.0
				MADX	−19.0	88.4	94.6
				MIL	−23.2	98.3	93.4
				TEX	−96.6	102.1	93.0

*The minus signs indicate acquisition later than predicted.

Table 2-4. Statistical Summary of USB Tracking Performance by Station During AS-506 (cont)

	TLC				TEC and Reentry		
Station	PAOS-AOS (Sec)*	% of Act/Pred Track	% of Valid/ Total Track	Station	PAOS-AOS (Sec)*	% of Act/Pred Track	% of Valid/ Total Track
ACN	-89.3	94.5	91.9	ACN	+3.0	96.3	99.5
ANG	-32.5	98.0	86.6	ANG	**	105.4	61.7
BDA	+194.0	101.0	93.5	BDA	+172.0	100.8	93.8
CRO	+149.3	100.6	94.6	CRO	+98.5	78.5	54.8
CYI	-140.5	93.4	87.6	GDS	-1.0	87.8	92.3
GDS	-3.0	89.2	94.6	GDSX	**	95.0	76.9
GDSX	-1.0	90.9	55.9	GWM	+88.0	90.3	94.9
GWM	-156.0	99.2	96.6	GYM	+31.0	108.9	86.2
GYM	-97.5	95.5	84.6	HAW	+49.7	100.0	81.2
HAW	-19.5	102.5	98.3	HSK	+96.7	93.0	84.3
HSK	-311.0	97.7	94.1	HSKX	-1.5	98.5	99.3
HSKX	-213.0	98.8	93.8	HTV	-93.0	32.0	0
MAD	**	76.0	91.5	MAD	**	98.5	94.5
MADX	-352.0	94.5	91.6	MADX	**	96.3	91.5
MIL	-66.8	100.0	91.5	MIL	**	108.5	95.1
TEX	-79.3	98.3	92.1	RED	+100.0	110.0	37.2

*The minus signs indicate acquisition later than predicted.

**All points deviate by more than 400 seconds.

(4) Tracking Performance During Lunar Stay. Table 2-5 shows the tracking performance of the stations tracking the LM during the time it was on the lunar surface. With the exception of ACN, all of the stations had only one view period which lasted from 2 to 12 hours.

2.2.1.3 Historical Data. Tables 2-6 and 2-7 and figures 2-3 through 2-8 present the data discussed above in comparison with data from previous lunar missions. Due to the dissimilarity between lunar missions and earth orbital missions, comparison between the two is inadvisable. Therefore, only the lunar missions are presented for comparison.

Table 2-6 shows the USB performance for all three lunar missions. These statistics show great consistency in the overall performance for these missions, with the only significant difference being in the SD for percentage of actual to predicted time between AS-503 and the other lunar missions. The differences in the other statistics are negligible. Figures 2-3 and 2-4 are graphic representations of the data contained in table 2-6.

Table 2-5. Tracking Performance During Lunar Stay

Station	Dev in Actual Acq to Pred Acq Time*	Percent of Actual Track to Pred View Time	Percent of Valid Track to Total Track Time
ACN			
1st View	-47	91.2	97.5
2nd View	+443	102.0	96.3
ANG	-10	95.7	82.5
CYI	-33	92.9	89.5
GDSX	+24	99.2	**35.5
HSK	+358	96.4	**37.9
MADX	-775	97.0	98.1

*The minus signs indicate acquisition later than predicted.

**Stations were configured for FM downlink and therefore valid track lines were not indicated. Consequently, these percentages do not reflect the quality of tracking by these stations.

Table 2-6. Statistical Summary of USB Tracking Performance During Lunar Missions

Mission	Percentage of Actual Track During Predicted View		Percentage of Valid Track to Total Track	
	Mean	SD	Mean	SD
AS-503 Dec 21, 1968	98.5	9.7	*	*
AS-505 May 18, 1969	98.1	13.2	90.5	15.5
AS-506 July 16, 1969	97.1	13.3	89.2	15.3

*Data not computed for this mission.

Table 2-7 shows the USB performance during each of the four phases for AS-505 and AS-506. (These statistics were not calculated for AS-503.) The statistics for the first three phases are remarkably similar with the exception that valid to total track during EO shows an improvement for AS-506 over AS-505. Some significant differences in the statistics of the two missions during TEC are evident. The SD for actual track to predicted view during this phase is significantly larger for AS-506,

2-12

showing that there were quite a few more deviations from the average than during AS-505. Even more significant is the difference in the means of valid to total track. The difference shows more difficulty during AS-506. Also significant is the fact that for AS-505 the two coast phases were statistically similar. This is to be expected as these two phases are physically similar. During AS-506 however, the SD for actual-to-predicted track and the mean for valid-to-total track show lower quality tracking during TEC than during TLC.

Table 2-7. Statistical USB Performance During AS-505
and AS-506 - by Phase

Mission	Percentage of Actual Track to Predicted View		Percentage of Valid Track to Total Track	
	Mean	SD	Mean	SD
Launch and EO				
AS-505	103.1	18.6	71.6	22.8
AS-506	103.7	23.2	79.2	14.3
TLC				
AS-505	97.6	7.6	93.0	9.1
AS-506	95.6	8.7	90.2	11.3
Lunar Phase				
AS-505	98.0	14.1	90.8	15.4
AS-506	97.4	12.1	90.9	14.8
TEC and Reentry				
AS-505	96.8	7.5	92.4	13.9
AS-506	94.9	17.8	85.2	17.5

Figure 2-5 gives a more detailed representation of the distribution of actual track to predicted view data points derived for the three lunar missions. The graph shows that AS-506 had a higher percentage of points below 90 percent than the others. It also had fewer points exceeding 110 percent than AS-503, but more than AS-505. The following example will illustrate how to read this graph: In order to find out what percentage of the AS-503 distribution falls between 90 and 110 percent, mark the point on the AS-503 line where the value of actual to predicted view is equal to 90 percent. The value for this point on the Y-axis is about 9 percent. This means that 9 percent of the data points were below 90 percent. Now find the point on the AS-503 line which equals 110 percent on the X-axis. The value of this point on the Y-axis is approximately 92 percent. This means that 92 percent of the points are below 110 percent. The percent of points between 90 and 110 percent can be found by subtracting the percentage of reports below 90 percent from the percentage of points below 110 percent, or 92 - 9 = 83%. Furthermore, we know that since 92 percent of the reports are below 110 percent, 8 percent must be at or above 110 percent.

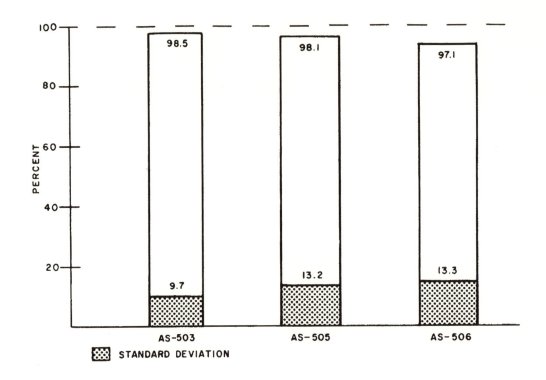

Figure 2-3. Percentage of Actual Track to Predicted View Time (USB)-- by Mission

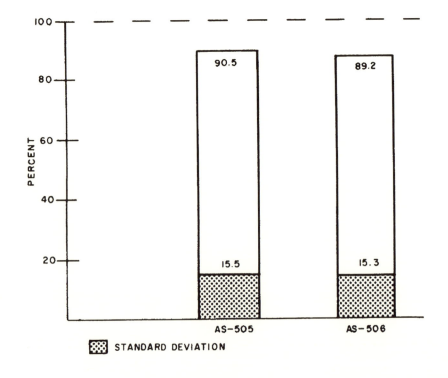

Figure 2-4. Percentage of Valid Track to Total Track (USB)--by Mission

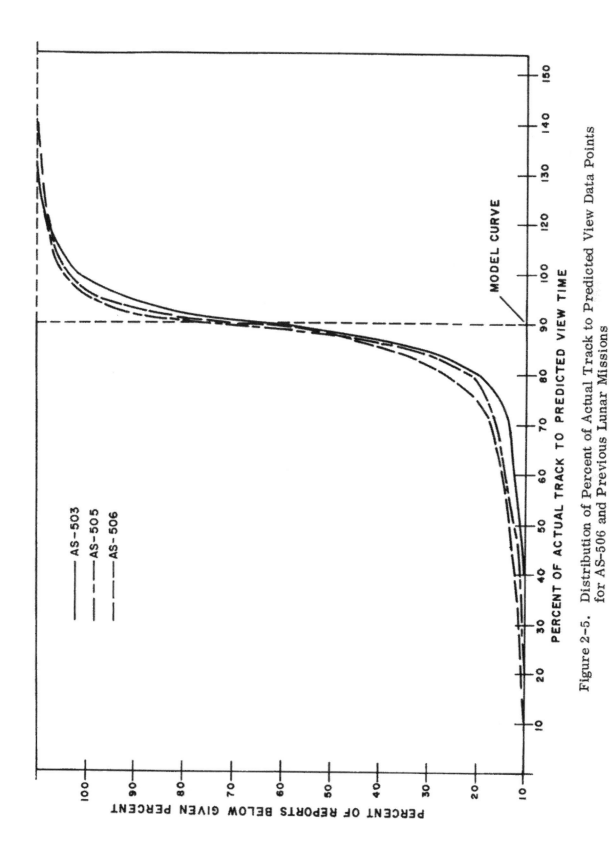

Figure 2-5. Distribution of Percent of Actual Track to Predicted View Data Points for AS-506 and Previous Lunar Missions

2-15

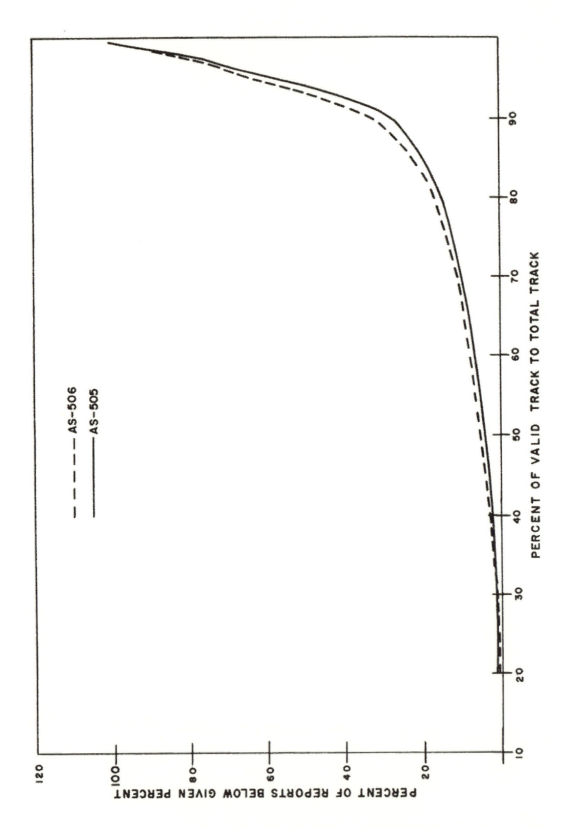

Figure 2-6. Distribution of Percent of Valid to Total Track Data Points for AS-506 and AS-505

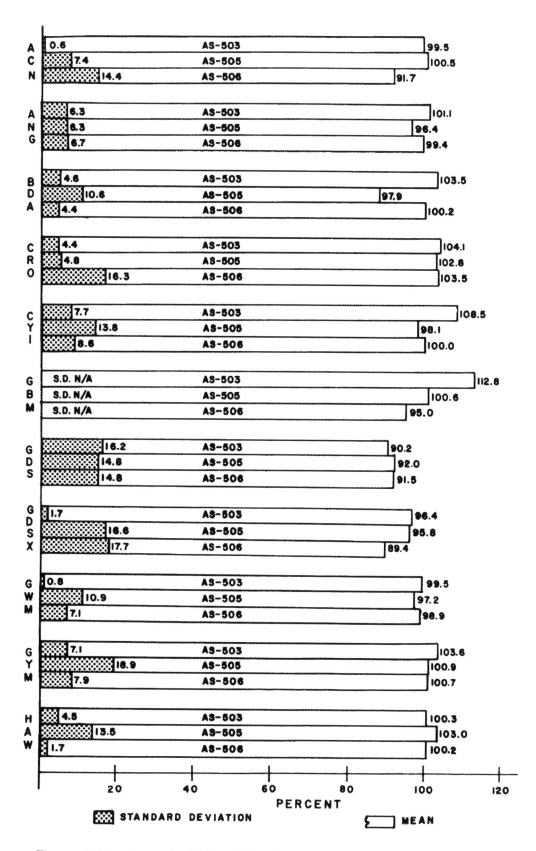

Figure 2-7. Percent of Actual Track to Predicted View Time--by Station

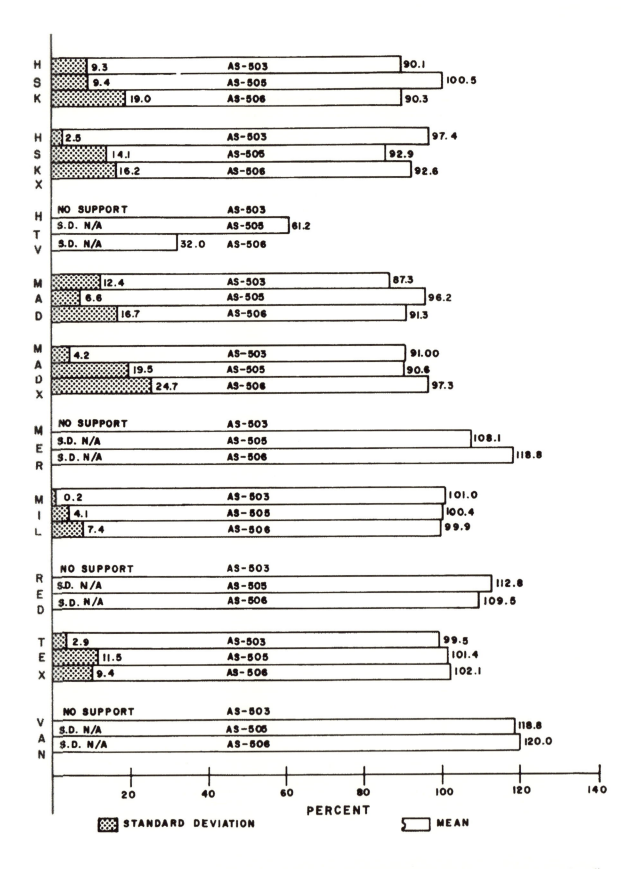

Figure 2-7. Percent of Actual Track to Predicted View Time--by Station (cont)

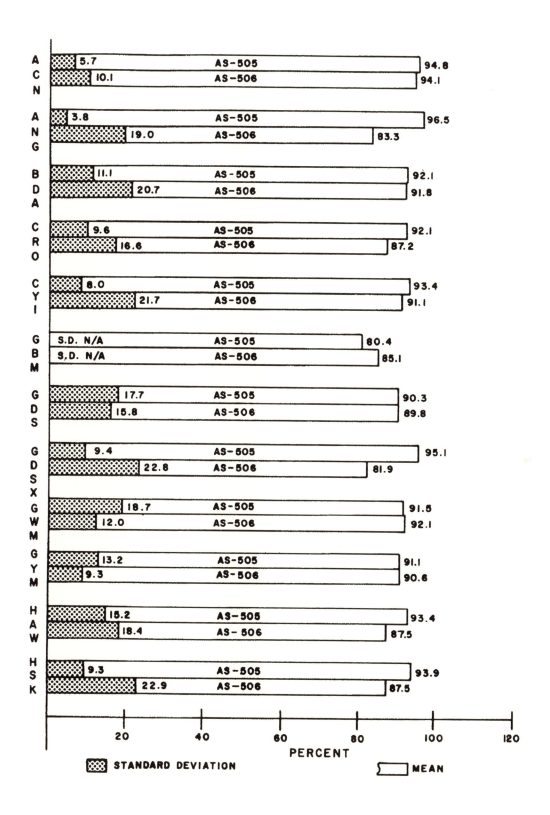

Figure 2-8. Percent of Valid Track to Total View Time -- by Station

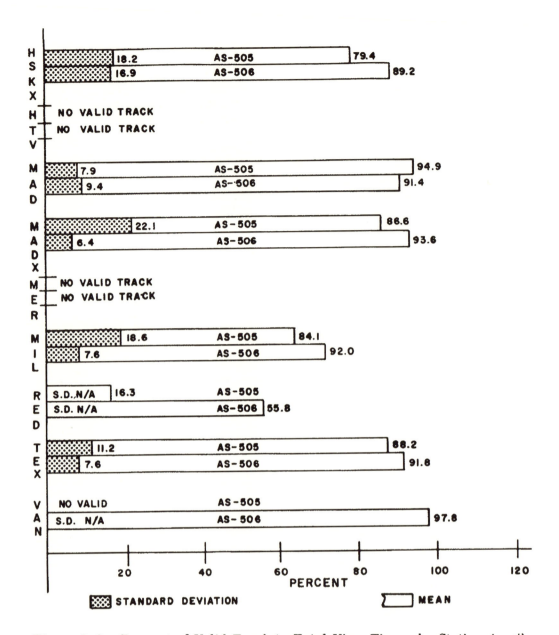

Figure 2-8. Percent of Valid Track to Total View Time--by Station (cont)

For AS-505, about 85 percent of the percentages fall between 90 and 110 percent, and for AS-506 approximately 77 percent of the points fall within this interval.

As can be seen by figure 2-5, the reason for the AS-503 SD being lower than the SD for other missions is that AS-503 did not include the extremely high and especially the extremely low points which are found in the other two missions.

Figure 2-6 shows the distribution of points used to compute the percentage of valid to total track for AS-506 and AS-505. This figure shows that AS-506 had more reports below 90 percent than did AS-505.

Figures 2-7 and 2-8 show the tracking performance by stations for the lunar missions. Figure 2-6 does not include AS-503, as valid to total track statistics were not computed for that mission.

2.2.2 C-BAND TRACKING

C-band radar met all mission requirements including range safety support, initial tracking data to establish early ephemeris, tracking the S-IVB/IU, and tracking the CM during reentry. A secondary mission requirement was not achieved when the RED received faulty SM acquisition information. This constituted the only support discrepancy.

The C-band tracking performance was analyzed statistically using the same three indicators which were used in analyzing the USB. The statistics are derived in the same manner as the comparable USB statistics, except that the radar PLIM's are the source of the actual track data. The GCC PLIM's are still the source of the predictions. Table 2-8 summarizes the three C-band tracking performance indicators for AS-506.

Table 2-8. C-band Statistical Summary for AS-506

Acquisition Deviation		Tracking Performance			
.Deviation Between Actual and Predicted AOS (Seconds)		Percentage of Actual Track During Pred View Time		Percentage of Valid Track During Total Track Time	
Mean	SD	Mean	SD	Mean	SD
–42.9*	78.4	98.8	21.9	88.6	15.6
*The minus sign indicates acquisition later than predicted.					

Figures 2-9 and 2-10 show the distribution of the individual data points used in calculating two of the statistics in table 2-8. Figure 2-9 shows that more than half of the reports indicated actual track times in excess of the predicted times, and almost 75 percent of the reports indicated actual track in excess of 90 percent of the predicted times. Figure 2-10 shows that for almost 70 percent of the reports, valid track was obtained for more than 90 percent of the total track time.

Table 2-9 and figures 2-11 and 2-12 offer a comparison of the C-band performance during the three lunar missions. Earth orbital missions are not included because the difference in view period and distance makes it impossible to meaningfully compare the tracking statistics.

Of the three indicators, percentage of valid track to total track is the best measure of tracking performance. In this respect, C-band performed better during AS-506 than during AS-505, but not quite so well as during AS-503. During AS-505, the aspect angle of the spacecraft antenna caused propagation problems. The aspect angle problem during AS-506 was not as great, and was at a minimum during AS-503. This factor could account for the statistical differences in valid to total track time. The performance comparison for the three missions, indicated by percentage of valid to total track, is also implied by the other two indicators.

2.2.3 VHF TRACKING

VHF was the primary acquisition source during EO and TLI. It was used for acquisition and tracking during these parts of the mission and performed exceptionally well.

2-21

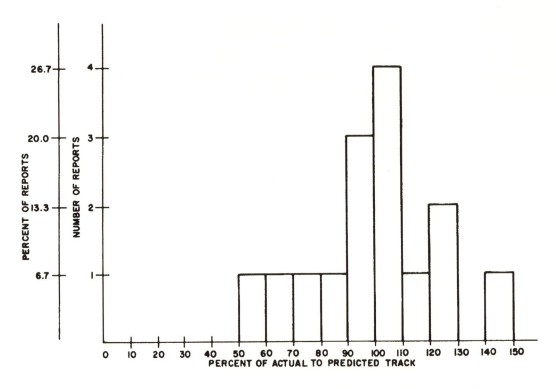

Figure 2-9. Percentage of C-band Actual Track to Predicted View

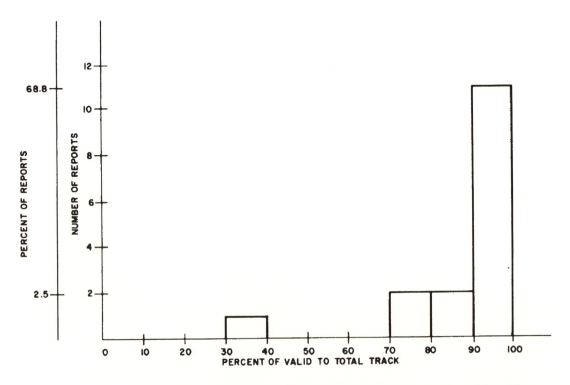

Figure 2-10. Percentage of Valid C-band Track During Total Tracking Time

Table 2-9. Statistical Summary of Radar Systems Performance
During Lunar Missions

Mission	Acquisition Deviation — Deviation Between Actual and Pred AOS (Seconds) Mean*	SD	Tracking Performance — Percentage of Actual Track During Pred View Time Mean	SD	Percentage of Valid Track During Total Track Time Mean	SD
AS-503 Dec 21, 1968	-13.2	23.4	97.9	10.1	94.2	6.5
AS-505 May 18, 1969	-45.4	102.1	72.8	30.4	84.1	21.5
AS-506 July 16, 1969	-42.9	78.4	98.8	21.9	88.6	15.6

*The minus signs indicate acquisition later than predicted.

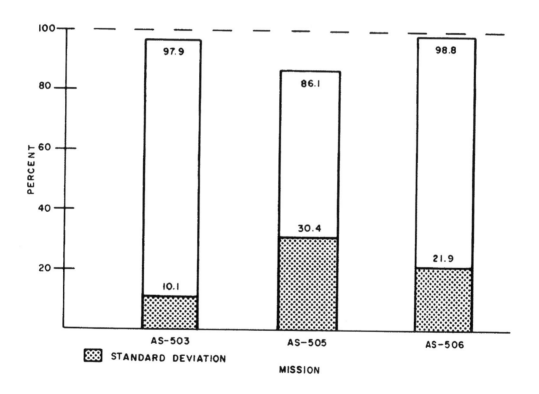

Figure 2-11. Comparison of Percentage of Actual Track to
Predicted View (C-band)--by Mission

2-23

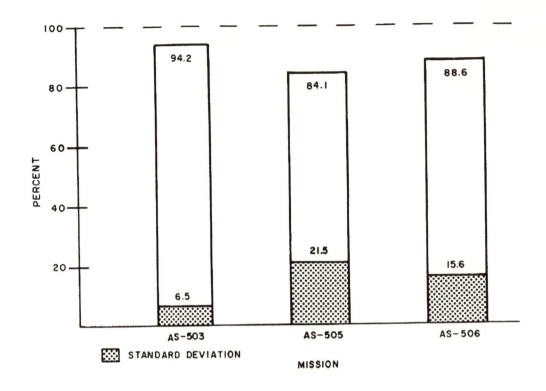

Figure 2-12. Comparison of Percentage of Valid Track to
Total Track (C-band)--by Mission

Table 2-10 summarizes the VHF tracking performance using two of the three indicators used for USB and C-band. The actual track data was obtained from the VHF PLIM's and the predictions were obtained from the GCC PLIM's. The data reduction was performed in the same manner as for USB and C-band. Percentage of actual track to predicted view was not included, as VHF was often used only as an acquisition source; therefore, this statistic would have no meaning as a measure of VHF performance.

Table 2-10. Statistical Summary of VHF Tracking Performance During AS-506

Deviation Between Predicted & Actual AOS (seconds)		Percent of Total Track Which Was Valid	
Mean	SD	Mean	SD
+16.8	37.3	90.8	11.0

2.3 TELEMETRY

Overall, the telemetry information received throughout AS-506 easily fulfilled mission requirements. This data was used by MCC and other agencies to monitor all systems of the space vehicles, as well as the physical condition of each of the astronauts.

Tables 2-11 through 2-14 give statistics for the telemetry performance from different perspectives. All of the statistics were computed from data listed on the PCM PLIM's. Total lock time was derived by taking the difference between the first lock time and the last lock time as listed on the PLIM. This time was then divided into the total dropout time (also listed on the PLIM) and the quotient was multiplied by 100. Thus was the percentage of dropout time to total lock time derived for each view period for each system at each station tracking during that view period. These percentages were then grouped in various categories and their averages and SD's were computed.

Table 2-11 shows the percentage of dropout time to total lock time by mission phase for USB telemetry. The percentage listed for a given phase includes all vehicles which were tracked during that phase.

Comparison of statistics among these phases must be tempered by consideration of the variation in the length of view periods during different phases as was discussed earlier for USB statistics. For example, a 10-percent dropout to total time for a coast phase may represent more than an hour of actual dropout time. This same 10 percent for an orbital view period during the lunar phase represents about 7 minutes of dropout time and less than a minute during an earth orbit. Consequently, a 10-percent average for EO represents better telemetry performance than it would during any other phase. During the orbital portion of the lunar phase, 10 percent would represent better performance than 10 percent during either of the coast phases, the only ones which can be directly compared.

Table 2-11. Percentage of Dropout Time to Total Lock Time--
By Mission Phases for USB Telemetry

Phase	Mean	SD
Launch & EO	6.6	10.2
TLC	19.4	30.4
Lunar Phase	11.0	22.6
TEC	11.8	19.1

Table 2-12. Percentage of Dropout Time to Total Lock Time--
By Vehicle for USB Telemetry

Vehicle	Mean	SD
CSM	12.8	24.3
LM	7.8	16.2
IU	4.6	11.8

Table 2-13. Percentage of Dropout Time to Total Lock Time--
By Tracking System

Tracking System	Mean	SD
USB	11.9	23.2
VHF	8.6	17.8

Table 2-14. Percentage of Dropout Time to Total Lock Time--
By Type of Station

Type of Station	Mean	SD
85-ft	2.9	5.5
30-ft with cooled paramps	5.3	13.2
30-ft without cooled paramps	24.1	31.2

Launch and EO have the lowest percentage of dropout time, as could be expected since the signal is stronger and more constant during this phase than during any other.

The largest percentage of dropout time occurred during TLC. During this phase, the HGA was not used as much as during the lunar phase and TEC. Since the percentage of dropout time is an inverse function of signal strength, it should be expected that the telemetry performance during TLC would therefore be inferior to that during the lunar phase and TEC.

It should be expected that the telemetry performance during the lunar phase would be better than during TEC, as the CSM was in the PTC mode during part of the latter and not at all during the former. At first glance, however, the statistics would seem to indicate essentially the same performance during these two phases. It must be remembered that, due to the differences in the length of view time between these two phases, similar statistics do not imply similarity of performance. The similarity of the statistics for these two phases does in fact imply a better performance during the lunar phase than during TEC. Also, two of the 30-foot stations which did not have cooled paramps, and therefore had a higher than average percent of dropout time, tracked during the lunar phase, but were released for ALSEP prior to TEC. This factor could be expected to result in some improvement in overall performance for TEC over the lunar phase, and at least partially compensate for the detrimental effect of occasional use of the PTC mode during TEC.

Table 2-12 presents the percentage of dropout time to total lock time by vehicle for the USB telemetry for the entire mission. The statistics show that IU telemetry was best. This is certainly due to the fact that the IU was tracked during EO or near earth when the signal was strongest.

The telemetry for the LM was not as good as for the IU due to the fact that the LM was being tracked only during the lunar phase and the signal was not as strong as during EO.

The telemetry for the LM was significantly better than that for the CSM. This difference can be attributed to several causes. The LM used only HGA, whereas the CSM used a low-gain antenna during part of the mission. Also, the CSM was being tracked during the coast phases when the tracking was inferior to the lunar phase, which was the only phase during which the LM was tracked.

The 85-foot stations had primary responsibility for tracking the LM. These stations averaged only 1.2 percent dropout time to total lock time, with an SD of 1.8. It is clear, therefore, that the average shown for the LM in this table is high primarily due to the high dropout rate of some of the 30-foot stations which were used for backup only.

Table 2-13 summarizes the performance of the USB and VHF telemetry for the entire mission. No comparison between the two should be made, as VHF was used only for

2-26

near-earth tracking and USB was used throughout the mission. Comparison can be made between the VHF statistics and those for EO statistics. Making this comparison, the USB telemetry was slightly better than the VHF.

Table 2-14 shows performance by type of station. Not surprisingly, the stations with 85-foot antennas averaged better than either type 30-foot stations and according to the SD also had far more consistent performance. Thirty-foot stations with cooled paramps averaged far better than 30-foot stations without them and with much greater consistency.

2.4 COMMAND

2.4.1 GENERAL

The overall performance of the MSFN command systems is described in the following paragraphs. The Range Safety Officer at Cape Kennedy had control of commanding activity during launch via the AFETR UHF command system. CSM/LM and S-IVB/IU commands were transmitted via the USB system.

2.4.2 COMMAND ACTIVITY

Table 2-15 shows the overall loading and command activity during the mission by station. The table is organized in the following manner:

a. Column 1. Number of loads sent to each station from MCC (total: 438).

b. Column 2. Total number of RTC executes by station (total: 3447).

c. Column 3. Total number of loads uplinked by station (total: 58).

d. Column 4. Total number of spacecraft rejects resulting from nonacceptance of uplinks by the spacecraft (total: 153).

e. Column 5. Total number of ground rejects resulting from improper station configuration and/or equipment malfunction (total: 48).

f. Column 6. Total number of uplink requests not received or not processed by the RSCC because of communications problems (total: 21).

g. Column 7. Total number of uplink requests not received or not processed because the RSCC was processing another uplink at the time and rejected the requests (total: 24).

h. Column 8. Total number of telemetry rejects resulting from incorrect downlist ID or no telemetry data received during uplinking of load data (total: 12).

i. Column 9. Total number of data rejects resulting from a nonvalid parity check between the uplinked command and the downlink telemetry return (total: 0).

2.4.3 USB COMMAND

The command support for AS-506 was very good. Few RSCC problems occurred during the mission. Only one occurred during a critical mission period (refer to paragraph 3.18.4). The total number of computer faults, spacecraft and ground rejects, and data losses was substantially reduced over the previous mission.

A total of 3885 command loads and RTC's were uplinked. During the mission, 48 ground rejects and 153 spacecraft rejects were received. A comparison of AS-505 reports shows a total of 68 ground rejects and 238 spacecraft rejects were received. The

2-27

Table 2-15. MSFN Loading and Command Activity

Station	(1) Loads Sent by MCC	(2) Total RTC's Executes	(3) Total Loads Uplinked	(4) S/C Rejects	(5) Ground Rejects	(6) Lost Executes	(7) RSCC Invac	(8) TLM Rejects	(9) Data Rejects
ANG	15	-	-	-	-	-	-	-	-
ACN	53	7	-	-	6	1	-	-	-
BDA	16	5	-	-	-	-	-	-	-
CYI	16	2	-	-	-	-	-	-	-
CRO	20	31	1	1	1	-	-	-	-
GDS	48	1196	16	11	20	5	14	4	-
GBM	15	-	-	-	-	-	-	-	-
GWM	28	110	3	4	2	-	-	-	-
GYM	15	-	-	-	-	-	-	-	-
HAW	35	19	-	3	-	-	-	-	-
HSK	39	891	8	52	6	9	3	2	-
MAD	56	1164	29	81	11	5	7	6	-
MER	15	-	-	-	-	-	-	-	-
MIL	19	2	-	-	-	-	-	-	-
RED	16	6	-	-	-	-	-	-	-
TEX	15	-	-	-	-	-	-	-	-
VAN	17	14	1	1	2	1	-	-	-
Mission Totals	438	3447	58	153	48	21	24	12	0

majority of the ground rejects resulted from attempts to execute during "unable to command" periods. The majority of the spacecraft rejects occurred during periods of marginal signal strength or by intentionally transmitting executes in the blind in order to reestablish communications. The Instrumentation Communications Officer was unable to command the spacecraft for a period of 4 minutes at 72:26:00 GET during TLC, and spacecraft rejects were received for all uplinks. The problem was cleared by the astronauts cycling the UPTELEMETRY/COMMAND RESET switch. A similar problem occurred during AS-504 and the cause is still unknown.

2.4.4 UHF COMMAND

The UHF command system performed nominally throughout the required support period. BDA transmitted the S-IVB safing command and received verification at 00:12:04 GET. All UHF command requirements were completed at 00:43:00 GET during EO revolution 1; the supporting stations were released from further support at 00:06:18 GET during TLC. No loss of UHF uplink data occurred during this mission.

2.5 AIR-TO-GROUND COMMUNICATIONS

A/G voice received at support stations throughout the mission was generally considered of excellent quality. A/G communications support met all requirements with a minimum number of equipment problems and operator errors. The following statistics are taken from the spacecraft communications PLIM data sheets and are based on signal quality received by the stations.

Table 2-16 summarizes spacecraft communications by configuration used and by quality of signal. It must be noted that quality is a matter of subjective judgment and is dependent upon the operator's ability to read the signal. During the entire mission, the quality was judged to be good or excellent 97.6 percent of the time that voice was used. The network used USB duplex configuration the entire time during the lunar and TEC phases and for all but the early portion of TLC. During EO, VHF simplex was used 40 percent of the time, VHF simplex and USB duplex 20 percent of the time, VHF duplex and USB duplex approximately 13 percent of the time, and USB duplex for the remainder.

Table 2-16. Quality and Configuration of
Spacecraft Communications

Configuration	Unreadable	Poor	Fair	Good	Excellent	No Voice Used	Total
Launch and EO							
VHF Simplex				1	4	1	6
VHF Duplex							
USB Duplex					1	1	2
VHF Simplex & USB Duplex				1	2		3
VHF Duplex & USB Duplex				1	3		4
Total				3	10	2	15
TLC							
VHF Simplex				1			1
VHF Duplex					1		1
USB Duplex			1	9	18	10	38
VHF Simplex & USB Duplex				1	4		5
VHF Duplex & USB Duplex							
Total			1	11	23	10	45
TEC							
VHF Simplex							
VHF Duplex							
USB Duplex				6	14	5	25
VHF Simplex & USB Duplex							
VHF Duplex & USB Duplex							
Total				6	14	5	25

Table 2-16. Quality and Configuration of
Spacecraft Communications (cont)

Configuration	Unreadable	Poor	Fair	Good	Excellent	No Voice Used	Total
Lunar Phase							
VHF Simplex							
VHF Duplex							
USB Duplex	1	1	2	54	86	36	180
VHF Simplex & USB Duplex							
VHF Duplex & USB Duplex							
Total	1	1	2	54	86	36	180
Total Mission							
VHF Simplex				2	4	1	7
VHF Duplex					1		1
USB Duplex	1	1	3	69	119	52	245
VHF Simplex & USB Duplex				2	6		8
VHF Duplex & USB Duplex				1	3		4
Total	1	1	3	74	133	53	265

2-31/2-32

SECTION 3. STATION PERFORMANCE

3.1 GENERAL

3.1.1 GENERAL PROBLEMS

Note

If a heading is not followed by text, this indicates that the
equipment covered by that heading supported without problems (refer to paragraphs 3.4.3.3, 3.9.5, and 3.12.5 for
examples).

3.1.1.1 GBM/BDA Handover. After launch at 00:06:00 GET, loss of signal was
experienced by BDA just prior to handover when GBM terminated the CSM uplink
carrier 30 seconds early. As a result, critical metric tracking data was briefly
interrupted prior to insertion into EO until BDA successfully reacquired. Analysis
of CSM stripcharts indicates that the SPE and the AGC were dropping out at the time
of IU handover because GBM had removed the CSM carrier at the same time the IU
was handed over from MIL to BDA. The SCM's called for the IU handover from MIL
to BDA at 00:06:00 GET and the CSM handover from GBM to BDA at 00:06:30 GET.

3.1.1.2 GDS Time Correlation Test. During the execution of time correlation tests
at 56:27:00 GET, the GDS ranging system was erroneously left in the clock mode as
opposed to the code modulation mode, making it impossible for three-way stations to
correlate time using two-way information, resulting in the tests being rescheduled.
Three-way stations affected by the error were ANG, BDA, GYM, HAW, MIL, and
TEX. The time correlation tests were run 10 minutes later with satisfactory results.

3.1.1.3 GDS/HAW Handover. An operator error during an unscheduled CSM HGA
handover between GDS and HAW on LO 3 resulted in a loss of approximately 6 minutes
and 30 seconds of active CSM support. At 80:36:00 GET, the GDS command computer
failed due to the failure of a diode in the primary data multiplexer cabinet, resulting
in a loss of command capability (3.10.4) and necessitating a handover to HAW at 80:40:18
GET. HAW initiated uplink at this time with GDS terminating uplink 10 seconds later.
However, the HAW R/E operator utilized omni antenna procedures as opposed to high-
gain antenna procedures. The HAW uplink swept through the spacecraft transponder
causing the HGA to slew off track, which resulted in a loss of downlink signal to all three-
way stations (ANG, ACN, BDA, CYI, GYM, MIL, and TEX). The HGA was manually
repositioned by the spacecraft crew approximately 6 minutes later. HAW then acquired
the spacecraft and tracked with no further problems.

3.1.1.4 Instrumentation Communication Officer Operator Error. At 151:57:00 GET, with
the spacecraft in the PTC mode of operation, the network had loss of signal from the
CSM due to the Instrumentation Communication Officer at MCC not switching antennas.
This loss of signal lasted for 48 minutes. GDS was the two-way station at the time.
Three-way stations affected were ANG, ACN, BDA, GDSX, GYM, MAD, and MIL.

3.1.1.5 Systems Monitor Recorders. Numerous systems monitor recorder problems
were encountered during this mission as they have been in the past. ANG reported
that the 100-channel event recorder failed at 153:53:00 GET (3.4.1). HSKX was unable
to operate recorder No. 4 in the "times 0.01" position (3.15.1). CYI, CRO, GWM, GYM,
HAW, and MIL all experienced problems with broken or faulty pens (3.8.1, 3.9.1,
3.12.1, 3.13.1, 3.14.1, and 3.19.1).

Additionally, CYI had a torn paper problem on the systems monitor, and the 100-channel
event recorder at MIL would not operate at a speed of 2mm/second.

3-1

3.1.2 USB

3.1.2.1 1218 Computers. The 1218 computers at ACN, BDA, CRO, GWM, and MADX experienced various problems during the mission. Although there was no significant mission impact, each incident impaired the program mode of operation for the USB antenna.

The BDA 1218 faulted while processing a 29-point acquisition messages at 81:15:00 GET (3.6.2.1). The program was reloaded and the message was processed properly.

During TLC, the operation of the 1218 at MADX was degraded due to the marginal operation of a high-speed tape punch (3.17.2.2).

The 1218 at ACN faulted intermittently during the last 3 days of the mission (3.5.2.1). The computer would not process acquisition messages nor could it be reloaded and processing be reinitialized.

At CRO, the 1218 faulted during the lunar phase and real-time APP tape drive was lost (3.9.2.1). GWM experienced 1218 problems during LO 18 and TEC (3.12.2.1). On both occasions the program was reloaded and normal operation resumed.

3.1.2.2 Power Amplifiers. Seven MSFN stations reported PA problems. Throughout the mission MADX had PA problems that were attributed to high ambient temperatures in the hydromechanical building (3.17.2.2). At TEX, a motor generator shutdown of the PA resulted in a loss of uplink capability at 25:35:00 GET (3.22.2.1). At 97:08:00 GET, ANG lost PA beam voltage because of an ac overcurrent interlock. The unit was recycled to restore it to operation (3.4.2.1). BDA experienced arcing of the beam voltage regulator during TEC which inhibited uplink capability (3.6.2.1). GDSX lost PA beam voltage on two occasions as described in paragraph 3.10.2.2. The first problem occurred while raising PA No. 1 power and the second problem resulted when the PA No. 2 body overcurrent relay tripped.

At 114:43:00 GET, the PA at GWM shutdown inhibiting uplink capability briefly. The shutdown resulted from heavy winds accompanying a local rain storm. Later, at 191:02:06 GET, GWM, which was the active two-way station at the time, again encountered a PA shutdown which was caused by a spark gap arc in the high-voltage power supply (3.12.2.1). HSKX experienced a major failure of PA No. 2 caused by a short circuit in the 400-volt, three-phase power supply. Considerable damage was sustained as detailed in paragraph 3.15.2.2.

3.1.2.3 Parametric Amplifiers. Three stations experienced problems with USB paramps. At 40:31:00 GET, CRO encountered paramp oscillations which caused momentary interruptions to tracking (3.9.2.1). HSK also experienced paramp oscillations during the lunar phase and TEC which adversely affected tracking as explained in paragraph 3.15.2.1. A similar situation occurred at MAD during LO 26 (3.17.2.1).

3.1.2.4 Precision Frequency Source. PFS problems occurred at two stations. MIL channel No. 1 experienced oscillations at 56:19:00 GET (3.19.2.1), while at MAD channel No. 1 failed at 98:08:30 GET (3.17.2.1). Neither problem had an adverse effect on mission support.

3.1.2.5 Flexure Monitor. Three of the four Apollo ships experienced failures of the USB flexure monitor. VAN reported a failure soon after departing port on July 7 (3.23.2.1). RED encountered a failure just prior to launch (3.20.2.1). MER's flexure monitor failed just prior to AOS during the TLI burn (3.18.2.1). All three ships exercised contingency procedures to provide support.

3.1.3 TELEMETRY

3.1.3.1 **CSM Received Signal.** All MSFN stations that supported USB downlink data streams incurred losses due to several common problems that are beyond the control of the stations. These problems affect the received signal and include signal fluctuations, spacecraft transmitting at an unfavorable aspect angle, changing bit rates of downlink telemetry, spacecraft antenna switching, terrain masking, and keyhole limitations.

A small data loss problem is still encountered by some MSFN stations during EO because of keyhole and terrain masking limitations. The keyhole problem at 30-foot stations has been greatly reduced over previous missions through the modification of antenna limitations in the Y-axis movement (EI-3282).

The most significant cause of telemetry data loss was the PTC mode of operation used by the spacecraft during TLC and TEC. In this mode of flight, the CSM revolves about its roll axis causing signal fluctuations from bad antenna aspect angles and masking of the transmitting spacecraft antenna from the ground station. The 30-foot stations were affected more than the 85-foot stations. This is attributable to the 10- to 15-dB increase in gain of the 85-foot systems over the 30-foot systems. As a result, the former were able to maintain decom lock whereas the latter dropped lock. An additional factor is a difference of decom lock capability between 30-foot stations with cooled paramps and those with uncooled paramps. The cooled paramp stations (GWM, CRO, HAW, ACN) have a 3- to 5-dB threshold improvement over the uncooled stations, which accounts for their better reception of telemetry.

Another cause of data loss was the bit rate changes of downlinked telemetry. When the bit rates were changed from low to high, the signal level decreased as much as 4 to 6 dB, which would cause a decom operating near threshold to drop out of lock until the signal reached threshold again. The MSFN decoms are able to maintain lock on LBR data (1.6 kbps) longer than on HBR data (51.2 or 71.2 kbps) due to a narrower bandwidth input filter being used. The 30-foot stations were primarily affected by this problem; those with uncooled paramps were affected more than those with cooled.

3.1.3.2 **LM Received Signal.** During the first 14 seconds of LM powered ascent from the lunar surface, all LM telemetry data was lost by stations supporting at this time (124:22:00 GET). This caused ACN, ANG, BDA, GWM, HSK, MAD, and MIL to loose decom lock. The loss of data was caused by the engine exhaust plume impinging upon the descent stage and/or lunar surface which resulted in a rapid phase change of the downlink PM data.

3.1.3.3 **Signal Conditioners.** Various problems with signal conditioners resulted in the loss of telemetry data at ACN (3.5.3.1) and MAD (3.17.3.1).

3.1.3.4 **Telemetry Computer.** During early TLC, both GWM and HAW experienced a fault which did not light the fault indicator. The problem occurred because a subroutine within the telemetry program executive routine was no longer in use and had been removed. However, the capability to call up this subroutine still existed within the program. Whenever this subroutine was called up, the program would fault without lighting the fault indicator. There was no mission impact as a result of this fault as HSK data was selected. Software modification has been implemented for AS-507 to resolve this condition.

A problem occurred at GWM, HAW, and MIL when the CADFISS test conductor was unable to remotely select the formats for H-70 CADFISS testing. Each time a remote telemetry selection was initiated, an error printout on the HSP would occur. Post-mission investigation of this problem revealed the cause to be in software at GRTC. The CADFISS remote telemetry format selection and corresponding high-speed error printouts occur when a NCG number other than 999 is used. No mission impact resulted. A postmission change to CADFISS telemetry procedures was implemented to correct this problem. Telemetry computer faults that occurred at ACN, BDA, GDS, GWM, HAW, MAD, and VAN are described in the respective station telemetry paragraphs.

3.1.4 COMMAND

Several USB malfunctions which affected command as well as other support capabilities were encountered. (Details of these problems can be found within paragraph 3.1.2.) No problems which affected UHF command were noted. Command computer malfunctions have all been included under the command heading of individual stations. These problems, stations affected, and reference paragraphs in parentheses are as follows:

3.1.4.1 Command Computer Faults. GDS (3.10.4), GYM (3.13.4), HSK (3.15.4), and MER (3.18.4) encountered command computer faults. Except for MER, automatic recoveries were accomplished successfully.

3.1.4.2 Command Computer Memory. ACN and BDA had memory core stack failures which were corrected by replacing the faulty stacks (3.5.4 and 3.6.4 respectively).

3.1.4.3 Command Magnetic Tape Unit Failure. MAD (3.17.4) and MER (3.18.4) experienced MTU failures which affected command capabilities.

3.1.5 AIR-TO-GROUND COMMUNICATIONS

Beginning at 114:04:13 GET during EVA, an echo was evident on the GOSS conference loop when the CapCom was uplinking to the LM. The problem appears to occur when both LM crew members are on the extra vehicular communications system. GDS and HSK confirmed that CapCom's voice was being received on the LM downlink. The MCC ComTech has stated that the echo problem is definitely inherent to the system and that high cost would be required to correct it. Houston is modifying equipment on the ground at MSC to alleviate the problem. A push-to-talk operated relay will open the CapCom receive path and therefore break the loop for the echo.

On numerous occasions the MCC ComTech contacted various stations in an attempt to determine the cause of distortion on GOSS conference. VOGAA operation is suspected as a contributing factor as reported by GDS and MAD, and the problem is being investigated by MFED. One modification of the VOGAA under consideration is to change the lag-time circuitry by use of larger capacitors and resistors, thereby increasing the time constant. Insertion of a bandpass filter to restrict passage of some noise frequencies and to permit more voice frequencies to pass is also proposed. In effect, these changes would slow down the development of attenuation when noise is present in the absence of voice. Then, on the resumption of voice, the VOGAA could respond to voice more quickly by not having so much attenuation to overcome. Implementation of these modifications for AS-507 is under consideration by MFED. Until a modification is approved, the problems with the VOGAA can best be approached under the existing configuration by directing the stations to verify that their VOGAA time-constant potentiometer is at maximum lag-time position, and that the EXPANSION control is reduced to provide approximately 15 dBm attenuation instead of the previous requirement for 35 dBm attenuation.

The VOGAA causing this distortion is somewhat questionable. Extensive tests were conducted under contract and the results revealed that in the expansion mode, inputs of -69 to +10 dBm resulted in a linear output relationship from -60 to -10 dBm. From

3-4

-10 to +10 dBm negative feedback was introduced. At no time during the tests was the output waveform distorted. The distortion reported by the stations was possibly the loss of the first part of the voice input following a period of attenuation of noise. It should be noted, however, that the simultaneous presence of voice and noise at the input of the VOGAA results in the presence of both in the output. This is inherent in the VOGAA original design. The proposed changes may alleviate these problems.

3.2 SPAN (Figure 3-1)

3.2.1 GENERAL

The purpose of SPAN is to provide data for a predictive service which will give advance warning of solar events which could endanger the lives of Apollo astronauts. The observable characteristics are related to the solar proton flux which follows these events. Solar events which occur during a manned mission are compared to those which have previously occurred so that a determination can be made of the possibility of a hazardous radiation environment.

SPAN facilities at CYI, CRO, and MSC participated in the mission. Each facility consists of an optical telescope in an observatory dome, a radio telescope in a radome, and a building containing photographic, recording, and communications equipment. (See figure 3-1.) Additionally, riometer systems at CYI, CRO, and Lima, Peru provided mission support.

Figure 3-1. A Typical SPAN Station

3.2.2 MISSION ACTIVITY

Solar activity was at a very low level throughout the mission. Activity was confined to subflares and a few "importance one" flares, all of which were small. The flares and subflares had no effect on mission activities. All messages and solar reports from SPAN facilities were received at MCC as required and no major problems were encountered.

Two minor equipment malfunctions occurred, but were resolved with little difficulty. At 25:13:00 GET, the radio telescope at CYI became inoperative because of a faulty data transmission multiplexer. Repairs were made and the unit was operational at 33:58:00 GET.

The radio telescope at CRO failed at 33:52:00 GET due to a broken shear pin in the antenna. The pin was replaced and the equipment was back in operation at 34:51:00 GET.

3.3 AIRA (Figures 3-2 through 3-4)

3.3.1 GENERAL

Eight aircraft (see figure 3-2) supported the AS-506 mission and at launch were positioned as follows: three at Darwin, Australia, one at Cocos, three at Guam, and one at Mauritius. This deployment provided maximum coverage for all possible launc azimuths.

The launch was nominal at 72 degrees azimuth and pre-TLI support was provided by ARIA's 3 and 4. ARIA 4 covered the required 2-minute TLI pre-ignition sequence start data interval. ARIA 3 recorded USB data with no dropouts, although VHF data during this period had some dropouts caused by multipath.

Figure 3-2. ARIA in Flight

3-6

3.3.2 MISSION PERFORMANCE

A recurring problem, that of the combiners not always selecting the clearest signals, appeared again on this mission. For example, during EO revolution 2 at 02:36:10 GET, the BER of the combined output dropped to the 10^{-2} level and remained at that level for 20 seconds. Had the combiner worked correctly, the BER should have increased after only 4 seconds (02:36:14 GET) when the data quality of the horizontally-polarized signal increased. Even so, the combiner performance has improved over that of previous missions.

During EO revolution 2 coverage there were four errors in the timing of the beam tilt function. The ARIA 3 antenna operator was 9 seconds late in changing from 16 to 11 degrees and from 11 to 0 degree when the antenna elevation was ascending. On the ARIA 4, the operator was 18 seconds early in making the change from 16 to 11 degrees while the antenna was ascending, and 8 seconds early in changing from 20 to 0 degree when the antenna was descending. All other changes were within 1.5 seconds of the scheduled time. Incorrect beam tilt caused some degradation of data from the horizontal polarization, but had no effect on the data from the vertical polarization.

After TLI, ARIA 6 and ARIA 8 rendezvoused with MER and RED respectively to receive in-flight transfer of the ships' TLI data. The data transfers were uneventful and these tapes, together with the data tapes from ARIA 3 and 4, were delivered to Huntsville, Alabama by ARIA 7 about 28 hours after TLI. (Figure 3-3 shows the geographical positioning for TLI support.)

Reentry was supported by ARIA's 3, 4, and 5. The uprange aircraft, ARIA 3, acquired prior to blackout and reacquired at blackout exit. The low quality data recorded after blackout exit was primarily due to range attenuation. There was a loss of approximately 2 minutes of data during the reentry period after ARIA 3 LOS until ARIA 4 and 5 AOS. Dropouts in ARIA 4 and 5 data were caused by the swinging and twisting of the spacecraft on the parachutes. While ARIA 5 had a data dropout at splash, ARIA 4 was able to record data. To ensure continuous telemetry coverage after the drogue parachutes are deployed, three aircraft should be deployed approximately 120 degrees apart around the splash point. (Figure 3-4 shows the ARIA positions for reentry support.)

The nominal launch azimuth did not require the use of ARIA 1, 6, 7, or 8 which were deployed for contingency coverage. These were released from mission status on July 18. ARIA 2 remained at Johnston Island to transport part of the lunar samples to Houston. Unforeseen delays on the U.S.S. Hornet resulted in the samples being delivered to Hickam AFB, where they were later picked up by ARIA 2 and delivered to Houston. ARIA's 2, 3, 4, and 5 were released from mission status on July 24. Table 3-1 summarizes the ARIA mission support.

3.4 ANTIGUA (ANG) (Figures 3-5 through 3-9)

3.4.1 GENERAL (Figures 3-5 through 3-7)

ANG provided passive support for TLC, the lunar phase, and TEC. During the lunar phase, the CSM was passively tracked for eight orbits and the LM for six orbits.

An operator error at GDS prevented correct time correlation tests at 56:27:00 GET (3.1.1.2).

While passively tracking the CSM on LO 3, a loss of downlink signal was encountered due to an unscheduled handover from GDS to HAW (3.1.1.3).

A 48-minute loss of CSM downlink occurred during TEC due to a procedural error at MCC (3.1.1.4).

3-7

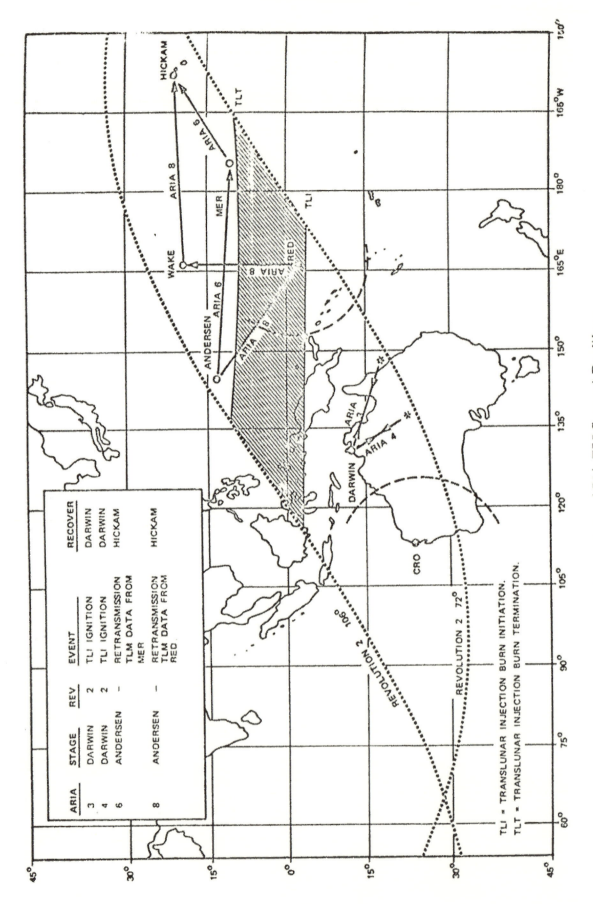

Figure 3-3. ARIA TLI Support Positions

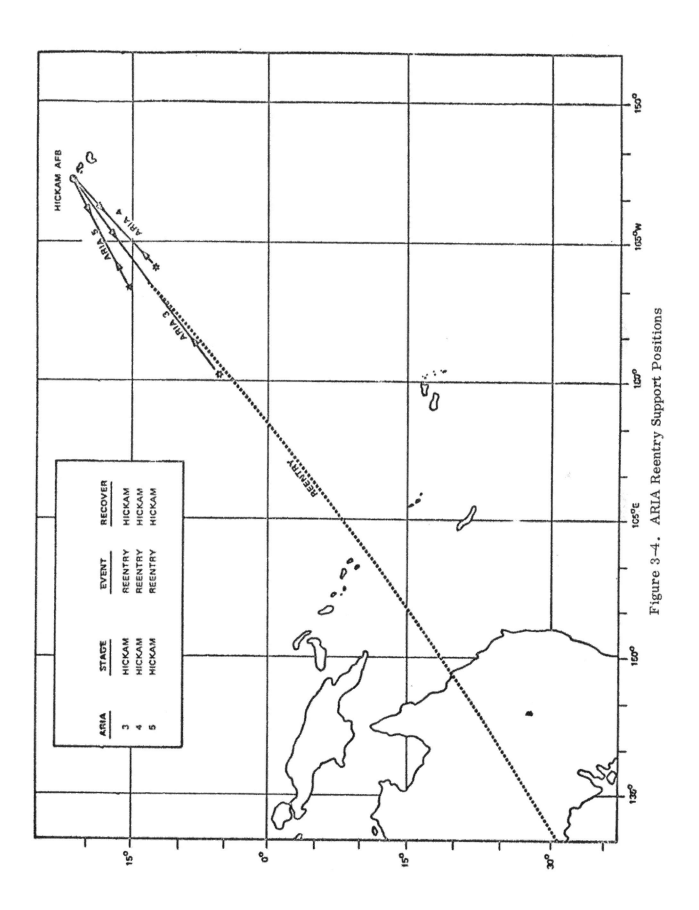

Figure 3-4. ARIA Reentry Support Positions

Table 3-1. ARIA Support Summary

ARIA No.	TSP Lat-Long	Time (GET) AOS	Remarks
TLI			
3	18°01'S 147°55'E	02:32:52	Good telemetry data. Excellent voice relay.
4	22°46'S 137°29'E	02:29:51	Good telemetry data. Satisfactory voice relay.
Reentry			
3	05°44'N 179°31'W	195:02:34	ARIA 3 acquired prior to and after blackout. Although not used by the spacecraft, voice relay was active and available. Telemetry data was good.
4	12°08'N 167°58'W	195:09:42	Voice relay was active and available to the space-craft. ARIA 4 stayed in the recovery area 1 hour and 20 minutes relaying the recovery operation to Houston. Telemetry data was good, but was affected by spacecraft maneuvers and its swings on its chutes.
5	15°10'N 169°44'W	195:09:30	Voice relay was active and available to the spacecraft. Telemetry data was good, but was affected by space-craft maneuvers and its swings on its chutes.

reported the 100-channel event recorder stopped running for approximately 20 ds at 153:53:00 GET. The recorder was restarted and no further problems wer ntered. The cause of the failure is unknown. Several stations encountered us problems with systems monitor recorders during the mission as discussed ir raph 3.1.1.5.

cedural problem was experienced beginning at 101:13:00 GET when the ANG Co receiving downlink from the spacecraft, but no uplink on Net 1. GSFC voice con alled to determine the status of Net 1. The ComTech was informed that false r being received and ANG had been removed from Net 1. The ANG ComTech had een previously informed of this action by GSFC voice control.

e were no telemetry or command computer problems.

Figure 3-5. ANG Tracking Station

Figure 3-6. ANG Mission Support

3.4.2 TRACKING

3.4.2.1 <u>USB (Figure 3-8)</u>. During TLC (59:58:00 GET), the optical collimation system became inoperative when operators were unable to focus the boresight TV camera. Investigation revealed a defective connector in the antenna feed-through panel. Replacement of the connector and realignment restored normal operation.

A PA failed because of ac overcurrent at 97:08:00 GET during the lunar phase and momentarily inhibited uplink capability. Recycling of the PA restored operation (3.1.2.2). At 99:18:00 GET, program mode tracking was interrupted for 4 minutes when 29-point acquisition message errors were transferred into the APP drive tape. When the autotrack mode of operation was selected, antenna movement was normal. The line feed function of the TDP was inhibited at 109:58:00 GET when PC card 2P48 failed. This defective card resulted in the loss of 4 minutes of LSD, although HSD remained normal. LSD was restored when the circuit card was replaced with a spare.

3.4.2.2 <u>Radar</u>. No requirements existed.

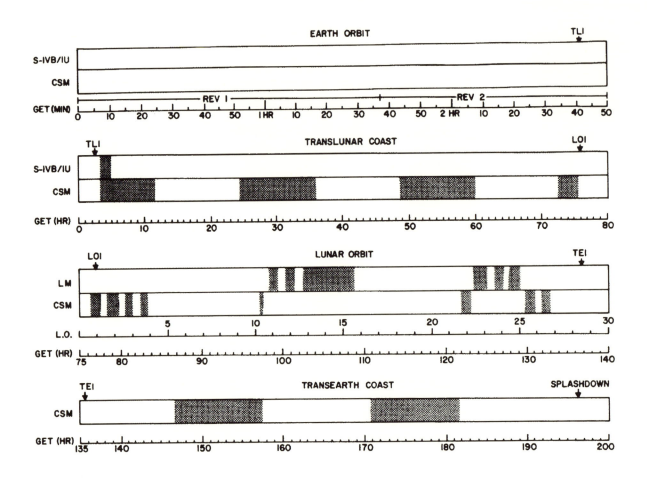

Figure 3-7. ANG Support Periods

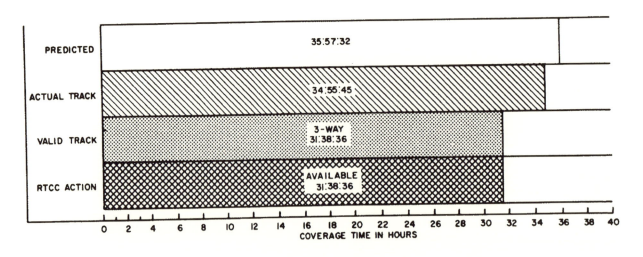

(a) TLC -- CSM

Figure 3-8. ANG USB Tracking Coverage

3-12

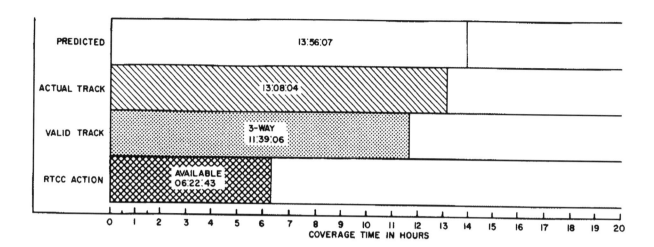

(b) Lunar Phase -- LM

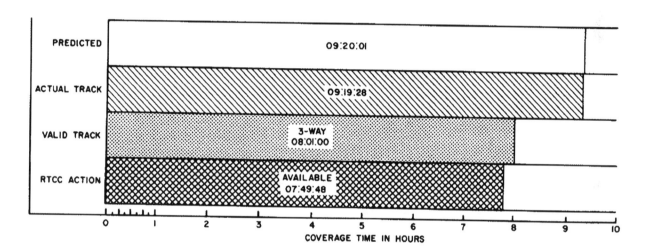

(c) Lunar Phase -- CSM

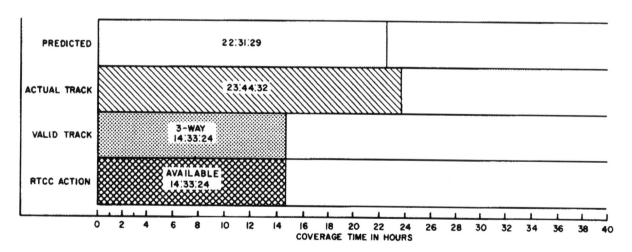

(d) TEC -- CSM

Figure 3-8. ANG USB Tracking Coverage (cont)

3-13

3.4.3 TELEMETRY (Figure 3-9)

3.4.3.1 USB. Real-time and dump data from the CSM and real-time data from the LM, after separation from the CSM, were received throughout the mission. There were periods when received data was substandard because of weak signals due to spacecraft attitude or its mode of flight (3.1.3.1). There were no significant equipment or operator problems. Approximately 14 seconds of LM data was lost during lunar liftoff due to degraded received signal (3.1.3.2).

3.4.3.2 VHF. The TELTRAC provided support for S-IVB and IU telemetry links simultaneously during TLC (03:04:10 to 04:53:16 GET). VHF tracking was used initially, then, as signal strength decreased with increased distance, the TELTRAC antenna was slaved to the USB to ensure VHF A/G support of the CSM. Decom lock was maintained on the S-IVB/IU link to a distance of approximately 20,700 nmi. There were some dropouts because of weak signals near LOS, which occurred shortly after the "slingshot" maneuver.

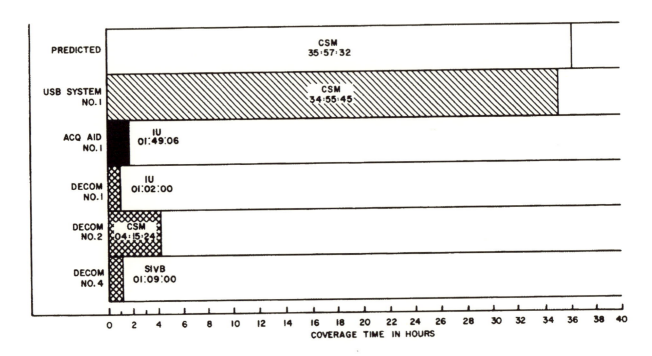

(a) TLC

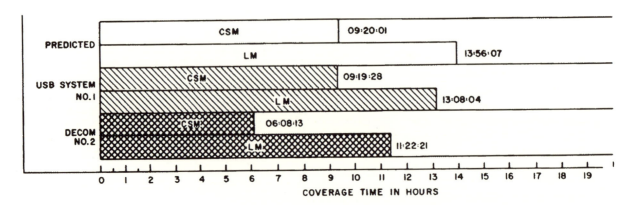

(b) Lunar Phase

Figure 3-9. ANG Telemetry Coverage

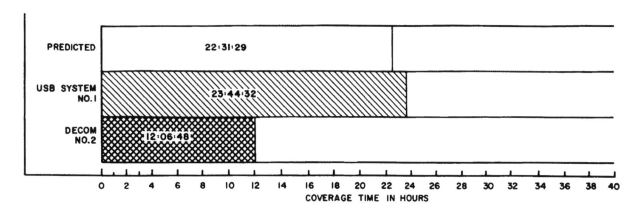

(c) TEC -- CSM

Figure 3-9. ANG Telemetry Coverage (cont)

3.4.3.3 Telemetry Computer

3.4.4 COMMAND

3.4.5 AIR-TO-GROUND COMMUNICATIONS

3.4.5.1 USB. Passive support was maintained with no requirement for remoting. Backup support was available for TLC, the lunar phase, and TEC.

3.4.5.2 VHF

3.5 ASCENSION (ACN) (Figures 3-10 through 3-14)

3.5.1 GENERAL (Figures 3-10 through 3-12)

The majority of the station's activities consisted of backup support for prime stations. However, over 3 hours of active two-way support of the LM was provided during the lunar surface operation. While passively tracking the CSM on LO 3, a loss of downlink signal was encountered due to an unscheduled handover from GDS to HAW (3.1.1.3). A 48-minute loss of CSM downlink occurred during TEC due to a procedural error at MCC (3.1.1.4).

No problems were encountered in A/G communications.

3.5.2 TRACKING

3.5.2.1 USB (Figure 3-13). ACN experienced no tracking problems until the lunar phase when the 1218 computer faulted intermittently. During LO 22 through LO 25, a total of 10 faults were observed when the CADCPS program would not process the 29-point acquisition messages properly. The program had to be reloaded after every three or four messages were processed. The computer continued to fault intermittently for the duration of the mission. Occasionally the computer would fault and could not be reloaded and at other times it would not process the acquisition messages properly. The failure did not impact mission support. The source of the problem is believed to have been the incompatibility of memory stacks with serial numbers ending in the digits 01 and 02. Subsequent to the installation of all stacks ending in serial number 01, ACN has had no unresolved problems (3.1.2.1).

3.5.2.2 Radar. No capability exists.

Figure 3-10. ACN Tracking Station

Figure 3-11. ACN Mission Support

3.5.3 TELEMETRY (Figure 3-14)

3.5.3.1 USB. S-band telemetry support was required during all of ACN's view periods during which CSM dump and real-time parameters, IU parameters, and LM lunar surface and orbital data were processed.

Minor losses of data were encountered by ACN throughout the mission due to signal fluctuations and changing bit rates. This problem was inherent throughout the MSFN (3.1.3.1). An operator error resulted in a 30-second data loss of a DSE dump playback to MCC during the lunar phase. A patch from the magnetic tape recorder output to the decom No. 2 signal conditioner was inadvertently removed at the DSDU. The decom was supporting the data dump playback, but data was not remoted to MCC while the patch was out. MAD was also receiving this data and was requested by MCC to remote CSM DSE data until ACN corrected their patching problem. Minor problems, as outlined in the following paragraphs, were encountered in the decom area, although data losses were insignificant due to the quick switchover to backup systems.

During NRT, decom No. 3 developed a history of chronic intermittent problems. Realignment of the wideband and narrowband signal conditioners seemed to resolve the

3-16

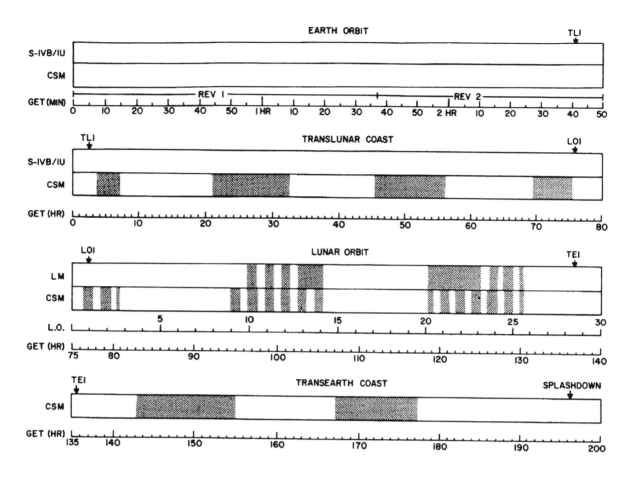

Figure 3-12. ACN Support Periods

problems each time. Apparently the reliability of the conditioners was marginal due to aging of components and became more critical after liftoff due to continuous use. At 13:50:00 GET the decom No. 3 memory failed to load or cycle properly, and gave a continuous parity error indication. The problem was apparently resolved by swapping PC cards in locations 11A1 and 12A1, and the system was operational at 15:53:00 GET. At 96:33:00 GET during LO 11, the decom No. 3 narrowband signal conditioner would not lock up on the CSM bit rates. The station reconfigured to support with the wideband signal conditioner, resulting in several seconds of data loss. The problem was traced to faulty resistors on the narrowband signal conditioner input circuitry. PC card 10B had two faulty resistors (OD-1 and OD-2) which had changed value causing the input gain to operate below threshold (3.1.3.3). Prime frame synchronization was not possible until the card was replaced and an alignment procedure completed. The problem was corrected and the narrowband signal conditioner became operational at 109:18:00 GET. However, during this period the wideband signal conditioner began dropping in and out of lock, requiring a minor adjustment of the bit rate VCO. Following all these corrective actions, the decom processed CSM HBR during the last 2 days of the mission with no reported data loss. Approximately 14 seconds of LM data was lost during lunar liftoff due to degraded received signal (3.1.3.2). Postmission diagnostics revealed no additional problems. There was no reliable telemetry data output from decom No. 3 during the first 5 days of the mission, but because decom No. 2 was configured to process the data simultaneously as backup, there was no data loss.

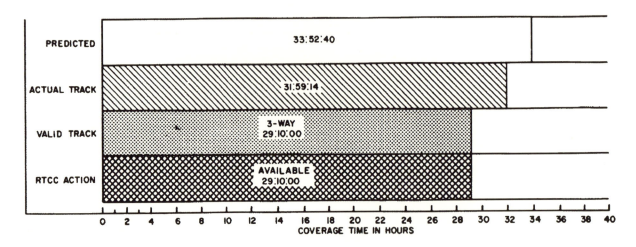

(1) TLC -- CSM

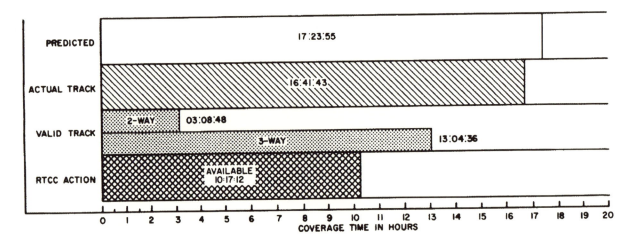

(2) Lunar Phase -- LM

(a) System No. 1

Figure 3-13. ACN USB Tracking Coverage

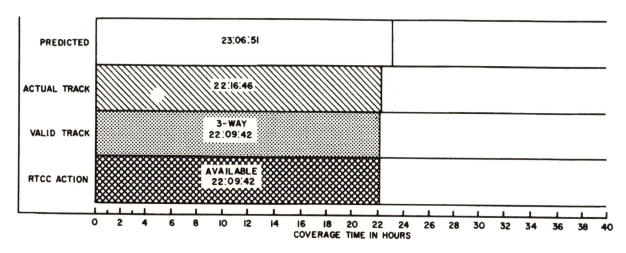

(3) TEC -- CSM

(a) System No. 1 (cont)

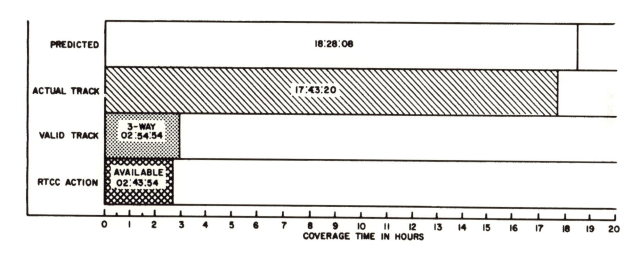

(b) System No. 2, Lunar Phase -- CSM

Figure 3-13. ACN USB Tracking Coverage (cont)

3.5.3.2 **VHF**. Support of IU and S-IVB data links was required only during the early part of TLC. During premission testing, the TELTRAC system was reported "red, can support," due to no output from the low-noise preamp right-hand circular polarization channel. Necessary parts were ordered and were received on July 15. Repairs were initiated immediately and the system was declared "green" at 1540 GMT the same day. The station provided its required support while slaved to the USB antenna system with no problems. Good signal quality was received with no loss of data.

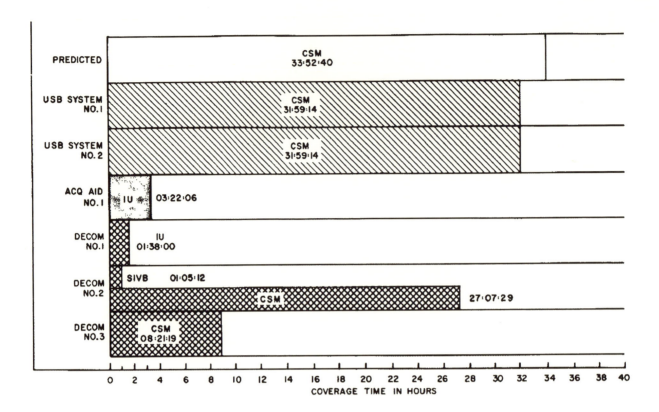

(a) TLC

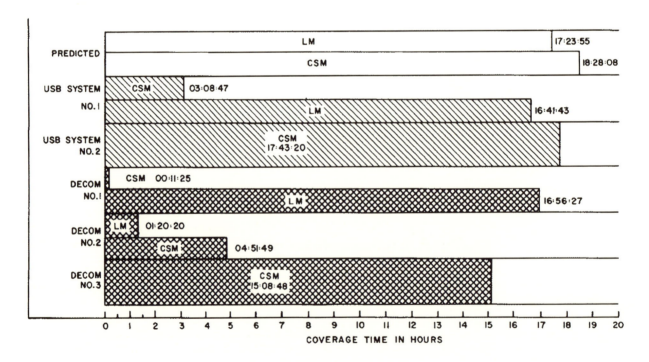

(b) Lunar Phase

Figure 3-14. ACN Telemetry Coverage

3-20

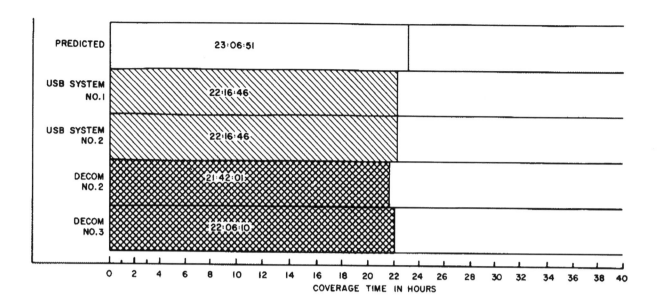

(o) TEC -- CSM

Figure 3-14. ACN Telemetry Coverage (cont)

3.5.3.3 <u>Telemetry Computer</u>. The 642B telemetry computer encountered five faults during the mission, all during TLC (3.1.3.4). Two faults occurred at 55:14:00 GET and 56:00:00 GET, respectively. In both instances, a FLTLD recovery was achieved and no data was lost. The computer faulted at 69:08:00 GET and FLTLD would not recover the program this time and a systems reload was required. No data was lost. About 1 hour later, at 70:16:00 GET, another fault occurred. The computer would not cycle the telemetry program and it was necessary to reverse the computer configuration to achieve support at this time. Approximately 10 minutes of real-time telemetry data was lost during this fault.

The computer faulted again at 75:33:00 GET. This had no impact on telemetry support as the computers were operating in a reversed configuration at the time. Maintenance on the computer revealed the source of the problems to be a poorly seated main memory stack. The unit was reported to be operational at 84:28:00 GET.

3.5.4 COMMAND

Prior to liftoff on launch day, the 642B computer designated for command was not operational due to a faulty memory core stack. The computer was dropping all bits in memory bank 1 (lower half of memory). The core stack was replaced before launch and the computer performed satisfactorily throughout the mission. (BDA had a similar problem, 3.1.4.2.) However, due to intermittent faults of the telemetry computer (3.5.3.3), the computers were operated in a reverse configuration during part of TLC. At 75:33:00 GET, the telemetry computer faulted while cycling the command program. However, CBARF recovery was successful and no data was lost.

At 76:27:00 GET during a contingency handover with MAD, ACN received six ground rejects for commands uplinked in a non-MAP override mode. The cause of these rejects was that modulation was being alternately applied and removed at 30-second intervals and the rejects occurred when commanding was attempted while the modulation was off.

3.5.5 AIR-TO-GROUND COMMUNICATIONS

3-21

3.6 BERMUDA (BDA) (Figures 3-15 through 3-21)

3.6.1 GENERAL (Figures 3-15 through 3-17)

BDA provided active tracking of the CSM and IU during launch and EO revolution 2. The remainder of the support periods consisted of passive track of the IU, CSM, and LM. While passively tracking the CSM at 06:00:00 GET just prior to handover from GBM, the downlink signal dropped out when GBM terminated the uplink early (3.1.1.1).

A/G communications were provided with no problems encountered.

An operator error at GDS prevented correct time correlation tests at 56:27:00 GET (3.1.1.2).

While passively tracking the CSM on LO 3, a loss of downlink signal was encountered due to an unscheduled handover from GDS to HAW (3.1.1.3).

A 48-minute loss of CSM downlink occurred during TEC due to a procedural error at MCC (3.1.1.4).

Figure 3-15. BDA Tracking Station

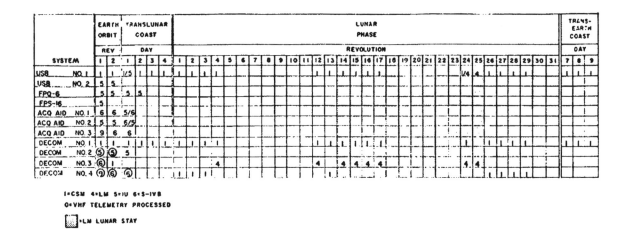

Figure 3-16. BDA Mission Support

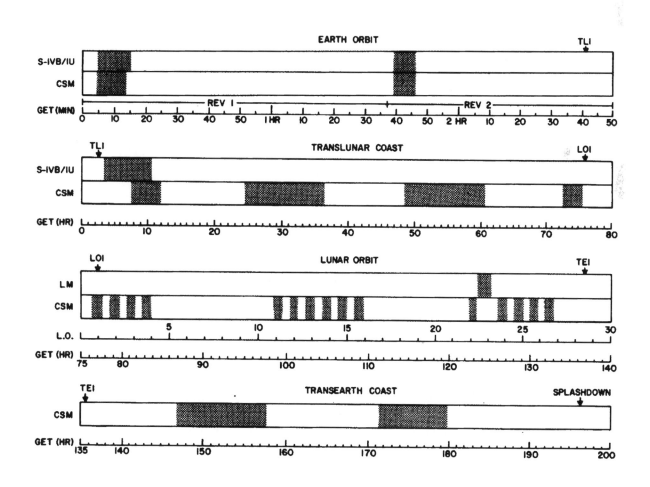

Figure 3-17. BDA Support Periods

3-23

3.6.2 TRACKING

3.6.2.1 USB (Figure 3-18). During the first day of TLC, the tracking data encoder power supply tripped off. This caused improperly encoded metric tracking information to be transmitted to GSFC for about 13 seconds until the circuit breaker was reset. The reason for the malfunction could not be determined.

During the same phase at 48:26:00 GET, a broken type hammer in the TDP printer caused a 20-minute loss of hard copy data. This data was recovered from reperforated tape after replacing the defective type hammer.

At the beginning of the lunar phase (81:15:00 GET), the 1218 computer faulted while processing a 29-point acquisition message. This caused a temporary interruption of the message processing sequence. The program was reloaded and processing of the message continued normally with no apparent reason for the fault (3.1.2.1). However, many hours of diagnostic testing indicated that the faulting was not caused by computer hardware.

Following CSM LO 24, Doppler data was invalidated for approximately 10 minutes after reconfiguring from CSM to LM tracking because the R/E operator failed to change the synthesizer to accommodate the LM frequency. The synthesizer was properly changed soon after the error was discovered. Fortunately, the station was supporting passively at the time and no mission impact resulted.

The station experienced internal arcing of the beam voltage regulator in the high-voltage power supply cabinet UD 10-10 during TEC at 145:13:00 GET (3.1.2.2). The problem was traced to a defective high-voltage relay (K-1). This inhibited uplink capability using the PA until the relay was replaced.

3.6.2.2 Radar (Figures 3-19 and 3-20). During launch, the FPQ-6 was required to provide C-band range safety tracking support and metric data for initial orbital determination. LTDS information was used as an acquisition source. Automatic tracking of the S-IVB/IU beacon was accomplished before the vehicle had reached 1 degree above the horizon. Strong beacon signals were received throughout the pass.

During EO revolution 2, the S-IVB/IU beacon was acquired by slaving to the FPS-16. This was necessary when an attempt to use an IRV failed to acquire the vehicle. At acquisition, the signal returns from the S-IVB/IU beacon were of poor quality. The operator's comment section of the PLIM's attributes this to interrogation of the two S-IVB/IU beacons simultaneously. At PCA, the signal quality had improved but 30 seconds of data were lost due to high azimuth rates when the vehicle passed overhead. Except for this 30-second dropout and delayed acquisition, the remainder of the pass was nominal. During TLC, the S-IVB/IU was beacon-tracked for over 8.5 hours to a range of more than 52,000 nmi. Although this range exceeds the design capabilities of the radar, it was possible to process the data and obtain range verification.

The FPS-16 was scheduled for backup support during launch and was slaved to the FPQ-6. During EO revolution 2, the FPS-16 provided support as an acquisition source for the FPQ-6. There were no tracking requirements other than this unscheduled support.

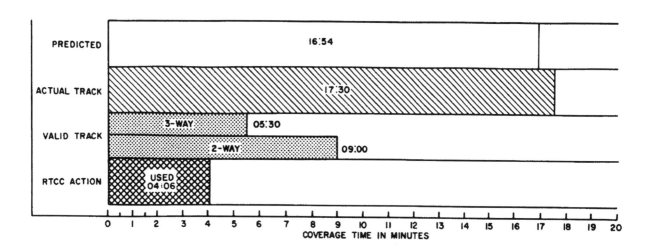

(1) EO -- CSM

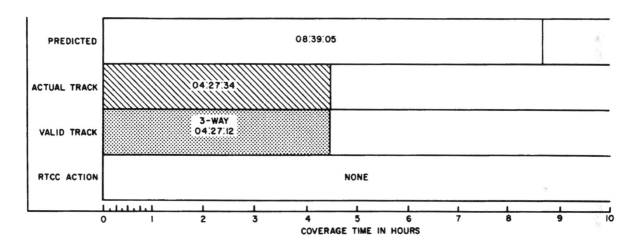

(2) TLC -- IU

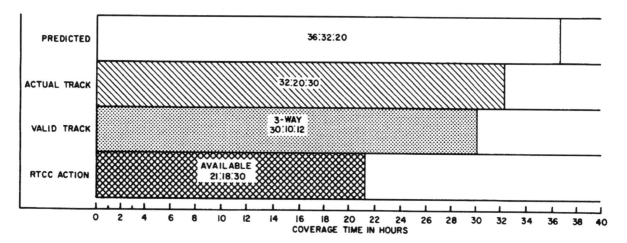

(3) TLC -- CSM

(a) System No. 1

Figure 3-18. BDA USB Tracking Coverage

3-25

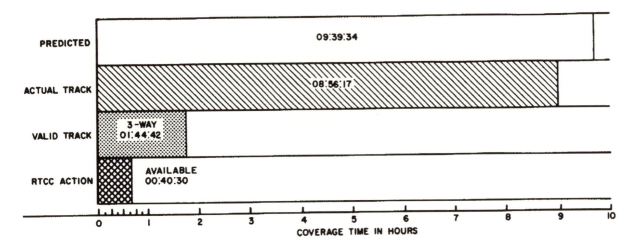

(4) Lunar Phase -- LM

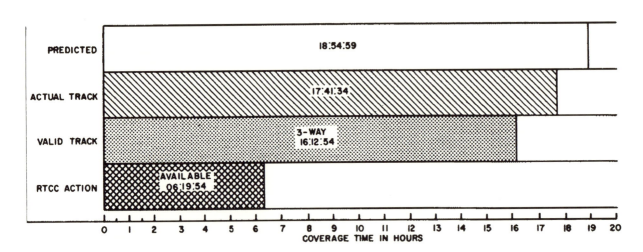

(5) Lunar Phase -- CSM

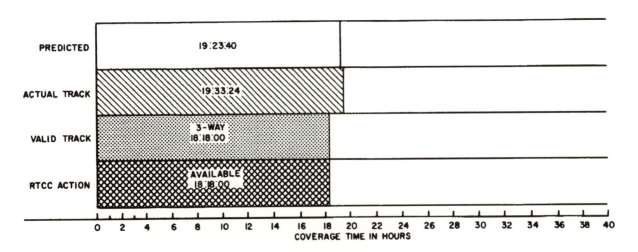

(6) TEC -- CSM

(a) System No. 1 (cont)

Figure 3-18. BDA USB Tracking Coverage (cont)

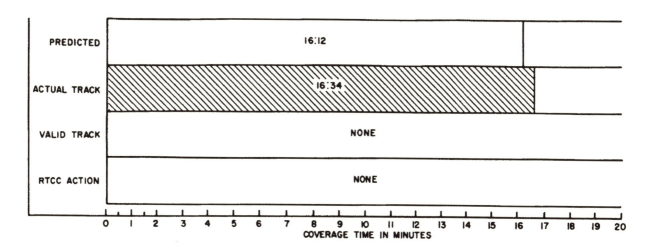

(b) System No. 2, EO--IU

Figure 3-18. BDA USB Tracking Coverage (cont)

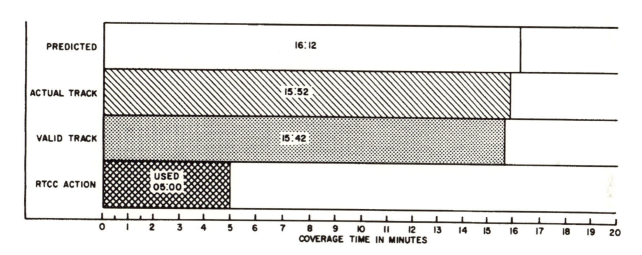

(a) EO -- IU

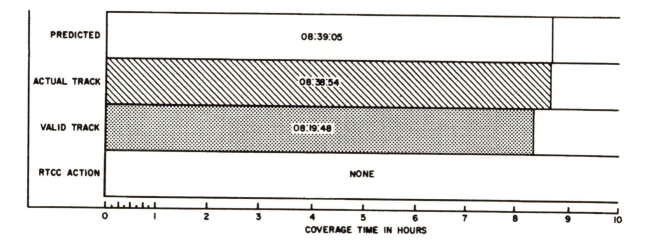

(b) TLC -- IU

Figure 3-19. BDA FPQ-6 Tracking Coverage

3-27

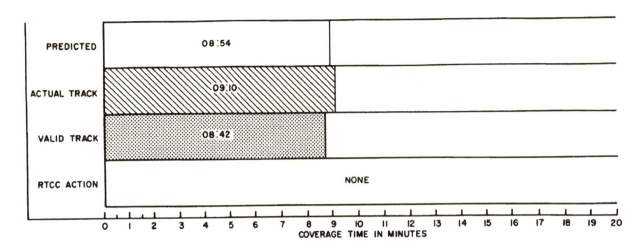

Figure 3-20. BDA FPS-16 Tracking Coverage, EO -- IU

3.6.3 TELEMETRY (Figure 3-21)

3.6.3.1 USB. Received signal strengths from the IU were good during launch and EO revolution 2, and decreased from fair to poor during TLC. Real-time and dump data from the CSM, and real-time data from the LM were received with a considerable amount of data losses noted during TLC, the lunar phase, and TEC. The losses were primarily a result of omni antenna usage on board the spacecraft and the PTC maneuvers (3.1.3.1).

Approximately 14 seconds of LM data was lost during lunar liftoff due to degraded received signal (3.1.3.2).

3.6.3.2 VHF. The three acq aid systems at BDA provided VHF telemetry support during launch, EO revolution 2, and the early portion of TLC. During TLC, received signal strengths decreased as the distance to the spacecraft increased. In addition, RFI with eventual loss of track was encountered during this phase by each acq aid whenever the antennas tracked the spacecraft into the sun. When this occurred, extreme solar noise was introduced into the receivers, degrading reception and causing loss of track. Reacquisition of the assigned vehicles was accomplished by slaving to another tracking system. Loss of signals caused by this phenomenon had no mission impact since the station was instructed to support passively as soon as the RFI was first encountered.

3.6.3.3 Telemetry Computer. During TLC (36:28:00 GET), the 1540 MTU No. 1 lost a vacuum motor on handler No. 2. The defective motor was replaced and the MTU was restored to operational status. BDA was on standby manning at the time; therefore, the problem had no mission impact. At 144:28:00 GET during TEC, the telemetry computer faulted while cycling the operational telemetry program. This occurred prior to Phase 3 of the SRT and automatic recovery was successfully achieved (3.1.3.4). Diagnostic tests indicated that bit 28 was failing in addresses 40000 to 77777. The current diverter cards were switched and apparently resolved the problem. To ensure against any recurrence, the current diverter card for bit 28 was replaced. The problem was corrected prior to processing commitments and did not affect mission support.

3.6.4 COMMAND

3.6.4.1 USB. During TLC (20:28:00 GET), the command computer experienced a memory problem. Bits 0 through 14 were intermittent in addresses 40000 to 77777. Extensive troubleshooting was performed and the problem was traced to memory core stack A2 in chassis A12. The defective memory core stack was severely loading the "Y" read current in the chassis, making all four stacks appear to be bad. Apparently a partial short in the stack caused a 100-ma loss of current. The defective stack was replaced and the system was restored to operational status at 45:28:00 GET. ACN also had a memory core stack failure (3.1.4.2). Uplink capability was affected during TEC (145:13:00) because of a defective relay (3.6.2.1). Neither of these problems hampered the mission, because the station was not required to uplink during this period. this period.

3.6.4.2 <u>UHF</u>. During launch, all range safety requirements were fulfilled. The safing command was successfully transmitted to the S-IVB at 00:12:04 GET after EO insertion.

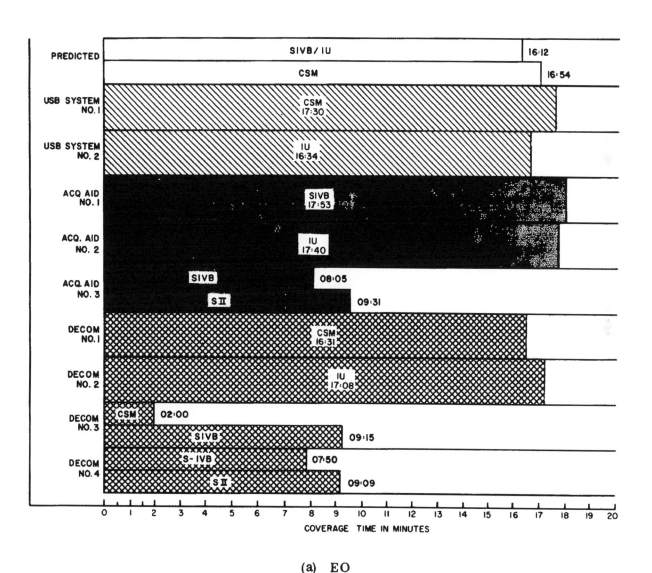

(a) EO

Figure 3-21. BDA Telemetry Coverage

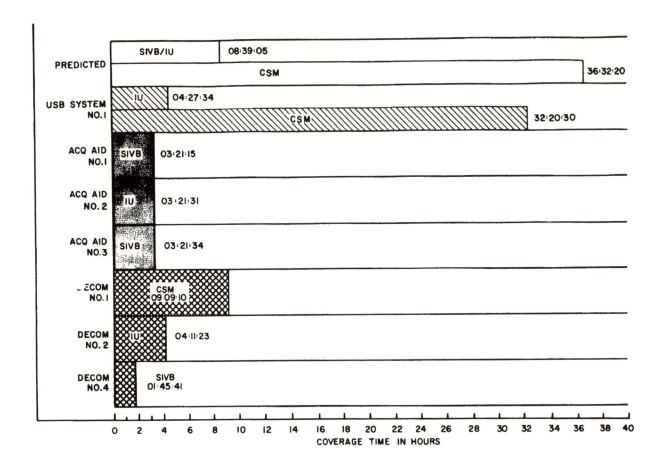

(b) TLC

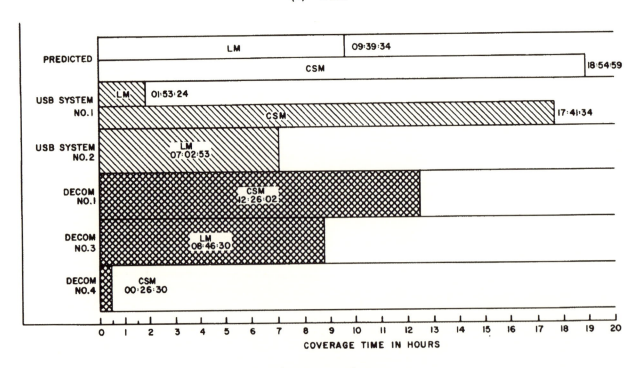

(c) Lunar Phase

Figure 3-21. BDA Telemetry Coverage (cont)

3-30

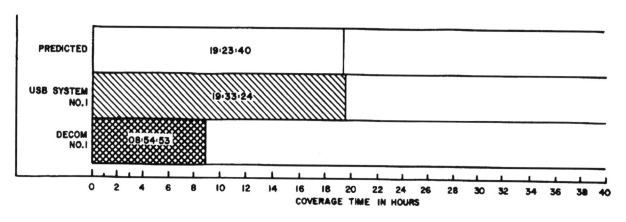

(d) TEC -- CSM

Figure 3-21. BDA Telemetry Coverage (cont)

3.6.5 AIR-TO-GROUND COMMUNICATIONS

3.7 **CALIFORNIA (CAL)** (Figures 3-22 through 3-25)

3.7.1 GENERAL (Figures 3-22 through 3-24)

CAL supported the S-IVB/IU on EO revolution 1 and the early portion of TLC, terminating activities at 08:05:50 GET. During these view periods the station also passively supported VHF A/G communications, with no problems encountered.

Figure 3-22. CAL Tracking Station

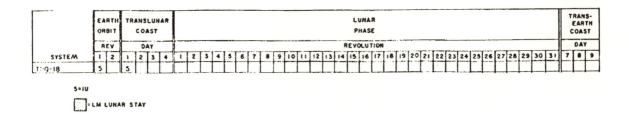

Figure 3-23. CAL Mission Support

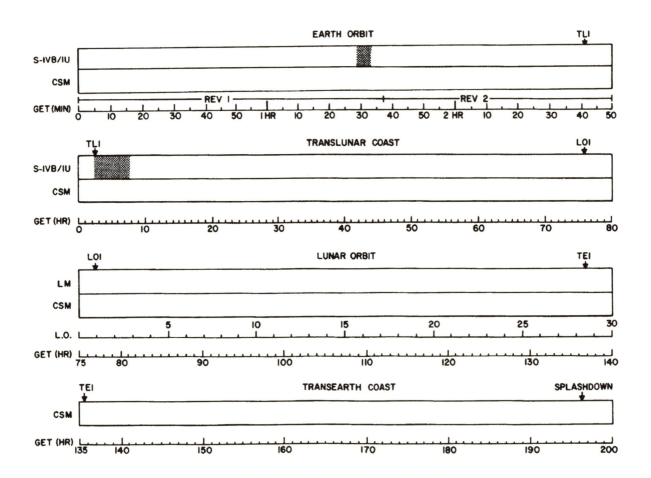

Figure 3-24. CAL Support Periods

3.7.2 TRACKING

3.7.2.1 USB. CAL has no USB tracking capability.

3.7.2.2 Radar (Figure 3-25). During EO revolution 1, acquisition of the IU was 1 minute and 32 seconds later than predicted. A multipath problem was experienced on this revolution that caused a data loss of 6 seconds. Initially, during this low-elevation pass, multipath signals from adjacent mountains caused the operator to track a sidelobe. The radar tracked the IU beacon in the early portion of TLC. The beacon was tracked for approximately 4 hours and 30 minutes; but the initial 1 hour and 30 minutes of CAL real-time data was invalid because an improper vehicle ID was being transmitted. However, this data was recorded and is being used for postflight analysis.

3-32

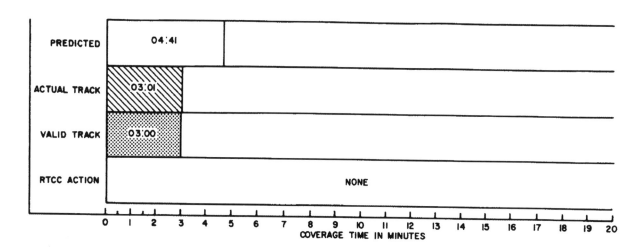

(a) EO -- IU

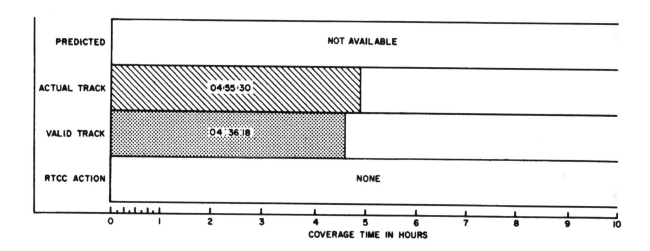

(b) TLC -- IU

Figure 3-25. CAL TPQ-18 Tracking Coverage

3.7.3 TELEMETRY

CAL has no telemetry processing capability.

3.7.4 COMMAND

No command capability exists at CAL.

3.7.5 AIR-TO-GROUND COMMUNICATIONS

3.8 CANARY ISLAND (CYI) (Figures 3-26 through 3-30)

3.8.1 GENERAL (Figures 3-26 through 3-28)

CYI provided active tracking of the IU during EO revolution 1 and the CSM during EO

revolution 2. Passive tracking of the CSM was provided during early TLC. Suppo
was shifted to the IU until the tracking requirement for this vehicle was terminate
Passive support of the CSM continued throughout the remainder of TLC. The CSM
passively supported during the lunar phase, after which passive support for the LI
was required until after touchdown on the lunar surface. When EASEP was deploy
by the LM crew, S-band at CYI was committed to support this requirement.

The 100-channel event recorder was halted for a paper change at 24:54:00 GET wh
passively tracking the CSM. Recorded events were lost until the paper change wa
completed. At 31:43:00 GET during this same view period, it was necessary to
deactivate channel 3 of the Mark 200 Brush recorder. A defect in the recorder pa
caused a tear in the channel 3 area and this pen was momentarily turned off until t
problem was resolved. No mission impact resulted because of these recorder pro
blems as the station was passively supporting. Other MSFN stations encountered
systems monitor problems (3.1.1.5). While passively tracking the CSM on LO 3,
loss of downlink signal was encountered due to an unscheduled handover from GDS
HAW (3.1.1.3).

The telemetry and command computers experienced no hardware or software prot
during required support periods.

A/G communication requirements were met with no difficulty.

Figure 3-26. CYI Tracking Station

3.8.2 TRACKING

3.8.2.1 USB (Figure 3-28). At 06:00:01 GET during TLC and after termination o
IU tracking, difficulty was encountered with the APP which resulted in unreliable p
gram mode tracking. The problem was apparently in the search mode circuitry a

3-34

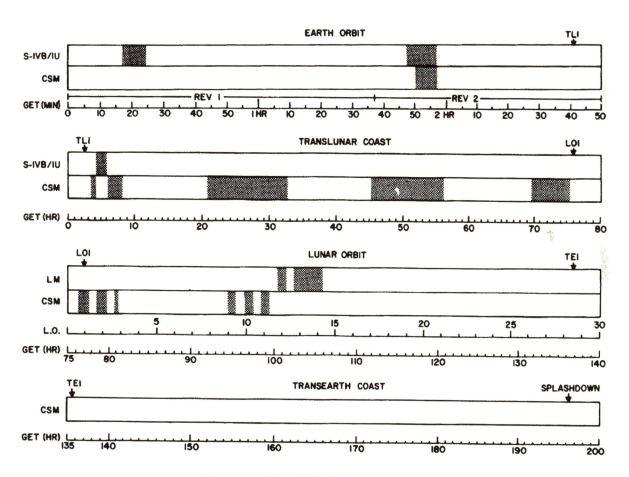

Figure 3-27. CYI Mission Support

Figure 3-28. CYI Support Periods

caused the 1218 computer to fault. No mission impact resulted because the system was passively autotracking the CSM at the time and this requirement was successfully accomplished.

During TLC (69:24:18 GET), the time interval counter of the range rate subsystem (used for the transmission of destruct Doppler data) failed for unknown reasons. Support of the mission was not affected because there was no requirement for this subsystem.

3.8.2.2 Radar. No requirements existed.

3-35

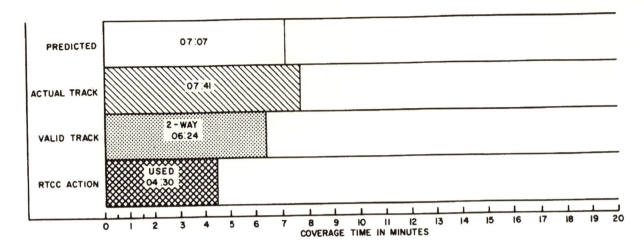

(a) EO -- IU

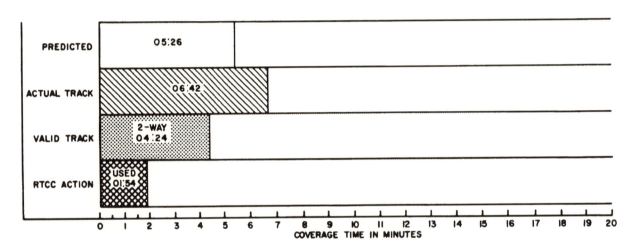

(b) EO -- CSM

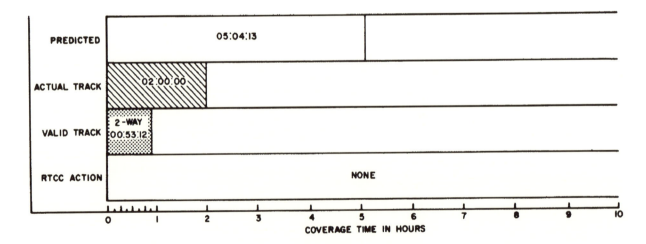

(c) TLC -- IU

Figure 3-29. CYI USB Tracking Coverage

3-36

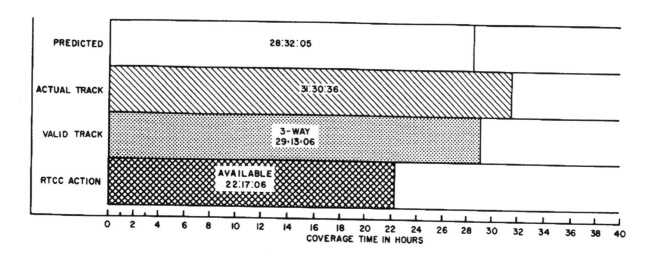

(d) TLC -- CSM

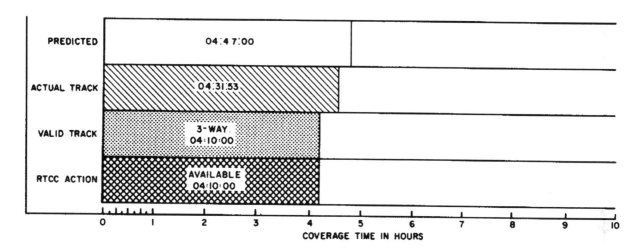

(e) Lunar Phase -- LM

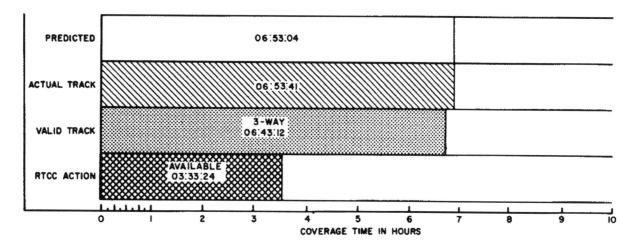

(f) Lunar Phase -- CSM

Figure 3-29. CYI USB Tracking Coverage (cont)

3.8.3 TELEMETRY (Figure 3-30)

3.8.3.1 USB. Support of the IU was provided during EO and early TLC. The CSM was also supported during these same view periods and, in addition, was supported in lunar orbits 1, 3, 10, and 12. LM support was provided during LO's 13 and 14, and until after lunar touchdown. Apollo telemetry requirements were terminated and CYI was committed to support EASEP shortly after deployment was completed.

Data losses experienced during the support periods were attributed to terrain and keyhole masking, use of the omni antenna on the spacecraft, the PTC mode of flight, and unfavorable spacecraft aspect angle (3.1.3.1).

3.8.3.2 VHF. Five minutes before launch, acq aid No. 1 experienced an azimuth amplidyne failure. An amplidyne salvaged from an obsolete UHF command system was used as a replacement to restore operation. No impact resulted because there was no requirement for this system to support. Acq aid No. 2 supported during EO and early TLC; however, the latter support was not a mission requirement. CYI elected to passively support until cessation of all VHF requirements. Signals were satisfactory throughout EO, but after TLI, as the spacecraft range increased, signal strength decreased as system limitations were approached.

3.8.3.3 Telemetry Computer

3.8.4 COMMAND

3.8.5 AIR-TO-GROUND COMMUNICATIONS

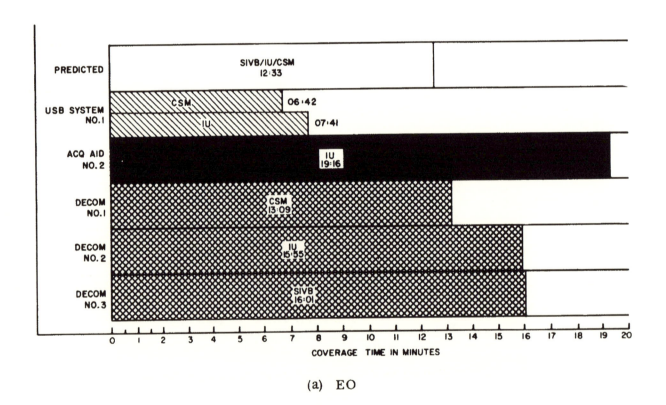

(a) EO

Figure 3-30. CYI Telemetry Coverage

3-38

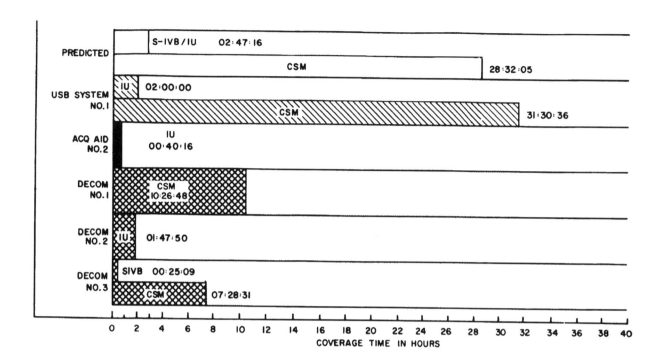

(b) TLC

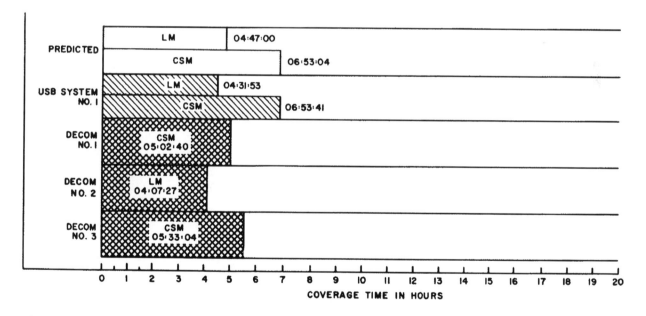

(c) Lunar Phase

Figure 3-30. CYI Telemetry Coverage (cont)

3-39

3.9 CARNARVON (CRO) (Figures 3-31 through 3-36)

3.9.1 GENERAL (Figures 3-31 through 3-33)

CRO USB System No. 1 actively supported the CSM during EO and the latter portion of TEC, and passively supported the CSM through TLC and seven lunar orbits. The LM was passively tracked for approximately 1 hour and 30 minutes during the latter portion of the lunar stay. System No. 2 actively supported the IU during EO and passively supported the CSM through TLC, the lunar phase, and TEC. The LM was passively supported by System No. 2 for approximately 21 minutes during LO 11 just prior to its undocking from the CSM.

During the lunar phase at 85:00:00 GET, pen A1 of the Brush 200 recorder failed and data recorded by this pen was interrupted. The pen was promptly replaced. This type of problem occurred at several MSFN stations (3.1.1.5).

CRO reported no A/G communication problems during the mission.

3.9.2 TRACKING

3.9.2.1 USB (Figure 3-34). During TLC, paramp oscillations caused momentary interruption to tracking. The paramp gain fluctuations were caused by the output instability of the klystron pump which had become unstable after only a few hours of operation. There was no significant impact as CRO reduced the paramp gain 3 dB and continued support until LOS. After LOS, the unsatisfactory klystron was replaced (3.1.2.3).

Real-time APP tape drive was lost because of occasional 1218 memory defects during the lunar phase. This malfunction was attributed to a failed sense output amplifier. Replacement of the amplifier restored APP antenna real-time drive (3.1.2.1).

Figure 3-31. CRO Tracking Station

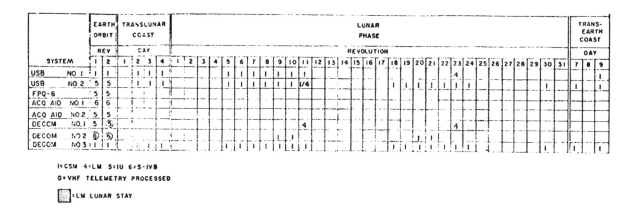

Figure 3-32. CRO Mission Support

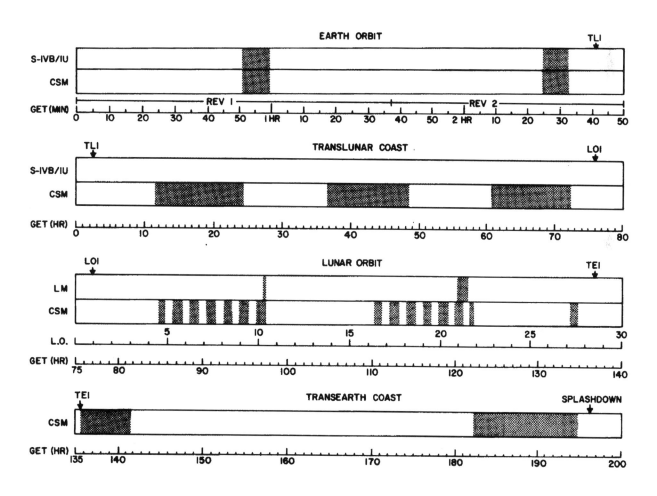

Figure 3-33. CRO Support Periods

3-41

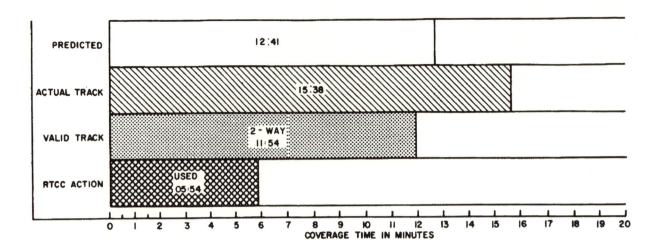

(1) EO -- CSM

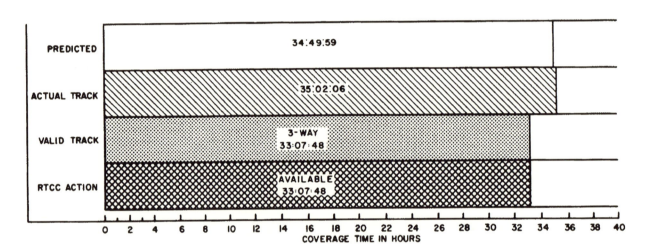

(2) TLC -- CSM

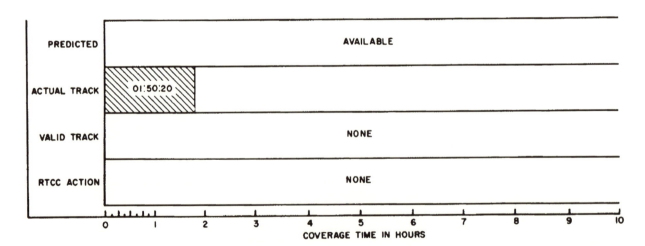

(3) Lunar Phase -- LM

(a) System No. 1

Figure 3-34. CRO USB Tracking Coverage

3-42

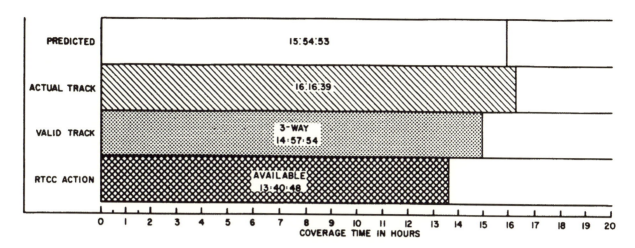

(4) Lunar Phase -- CSM

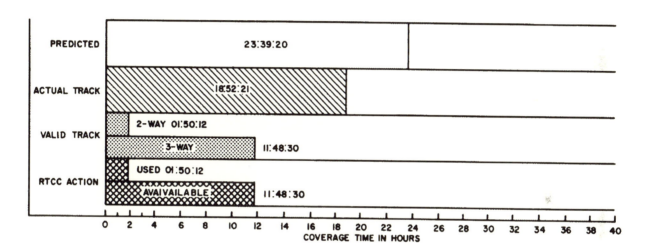

(5) TEC -- CSM

(a) System No. 1 (cont)

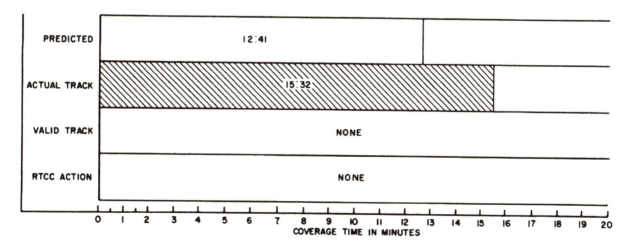

(b) System No. 2, EO -- IU

Figure 3-34. CRO USB Tracking Coverage (cont)

3-43

3.9.2.2 <u>Radar (Figure 3-35)</u>. During EO revolution 1, the FPQ-6 supported the IU. Acquisition was 32 seconds earlier than the predicted time, and the vehicle was tracked for 37 seconds longer than predicted. The IU beacon was acquired using an IRV, and the beacon signal level enabled tracking at an elevation of less than 0.5 degree. No data loss was incurred during this revolution.

EO revolution 2 was also a low-elevation pass; the IU beacon was acquired 22 seconds later than predicted. The duration of the actual track was 30 seconds less than the predicted view period. One hundred percent of the metric tracking data during this revolution was tagged valid.

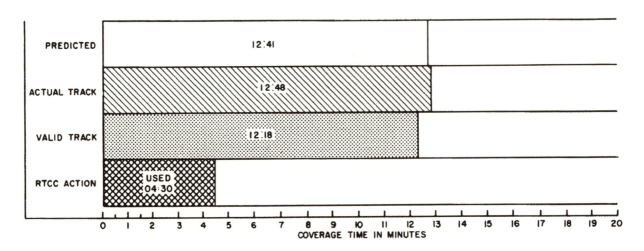

Figure 3-35. CRO FPQ-6 Tracking Coverage, EO -- IU

3.9.3 TELEMETRY (Figure 3-36)

3.9.3.1 <u>USB</u>. Real-time S-band telemetry support was provided for the CSM during EO, TLC, the lunar phase, and TEC. Support of the IU was provided during EO revolution 1, and LM telemetry was supported during LO 11 and EVA, with CSM dump data provided during the lunar phase. Momentary loss of data was experienced because of fluctuating signals, but had no significant impact on the mission. Overall telemetry support was good, and only at those times when the spacecraft was using the omni antenna or the PTC mode of flight did minor processing problems occur (3.1.3.1). During the lunar stay at 120:26:00 GET, CRO was advised by MCC that LM data was not being received at CCATS. The station had been requested to remote this data in addition to supporting EASEP. The problem was resolved when CRO removed the computer buffer plugs inhibiting this data from being input to the telemetry computer for HSD format processing. No data losses occurred because CRO was only required to provide backup support.

3.9.3.2 <u>VHF</u>. Acq aids No. 1 and No. 2 (TELTRAC) supported VHF telemetry links from the S-IVB and IU respectively during EO. During TEC, both aids, slaved to the USB, received very good voice from the CSM at a distance of approximately 9,000 nmi. No remoting requirement was received from MCC as USB A/G voice was prime during this view period.

Acq aid No. 2 operated during the entire mission on a "red, can support" basis, using a temporary antenna cable configuration. There were no significant equipment or operator difficulties encountered with the acq aids during VHF telemetry support.

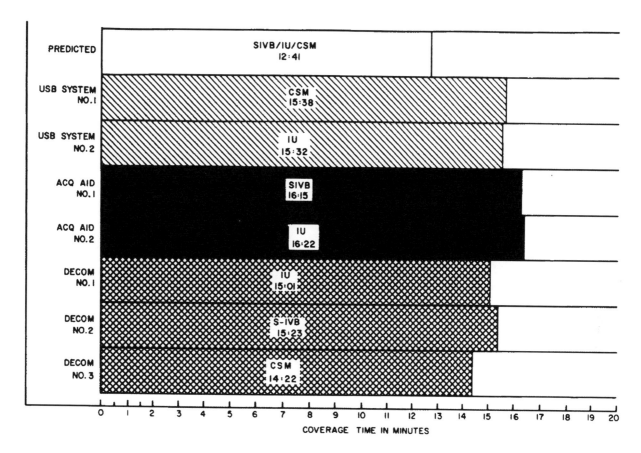

(a) EO

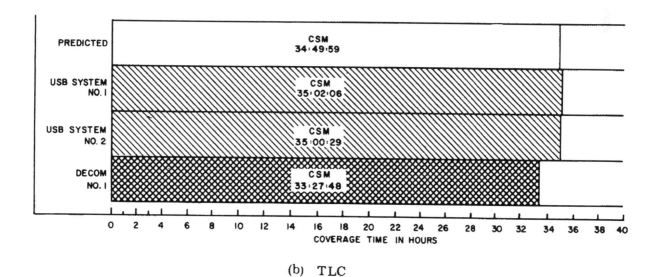

(b) TLC

Figure 3-36. CRO Telemetry Coverage

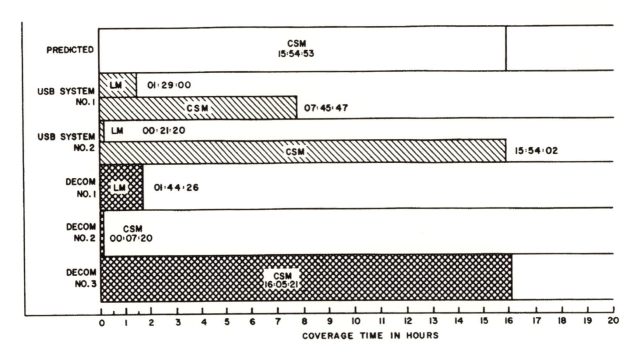

(c) Lunar Phase

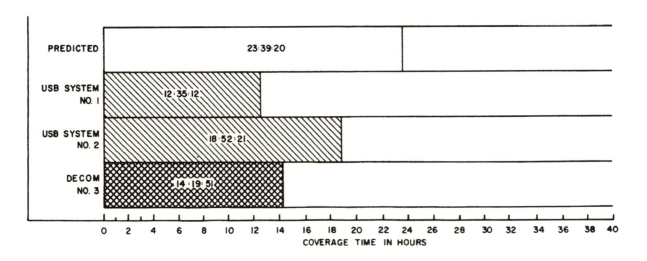

(d) TEC -- CSM

Figure 3-36. CRO Telemetry Coverage (cont)

3.9.3.3 <u>Telemetry Computer.</u> During three previous Apollo missions (AS-205, AS-504, and AS-505), the CRO telemetry computer faulted on numerous occasions. After AS-505, extensive investigation revealed that PC card C3 was temperature sensitive. The replacement of that card has evidently solved this problem, because there were no computer faults reported during this mission.

3.9.4 COMMAND

During the terminal count, CRO experienced difficulty with their S-band command interface test by uplinking an incorrect SLV RTC. A command review at CRO revealed that the data in the memory core for this command was incorrect. The command computer was reloaded and the interface test was completed successfully. The reason for this computer fault has not been determined. During TLC difficulty was experienced with command computer DTU No. 5. Troubleshooting revealed a failed PC card 23B. There was no significant mission impact because at this time GDS and HSK had spacecraft acquisition.

3.9.5 AIR-TO-GROUND COMMUNICATIONS

3.10 GOLDSTONE/GOLDSTONE WING (GDS/GDSX) (Figures 3-37 through 3-42)

3.10.1 GENERAL (Figure 3-37 through 3-39)

GDS actively supported EO revolution 1, TLC, the lunar phase, and TEC. GDSX supported EO revolution 1, TLC, the lunar phase including the LM lunar stay, and TEC. Limited active support was provided for the CSM and IU on EO revolution 1. Both active and passive support of the CSM and IU were provided while the vehicles were in TLC. After LM separation from the CSM in lunar orbit, active and passive support were provided for the LM in its descent to lunar touchdown, during the lunar stay, through its ascent from the lunar surface, during its rendezvous and docking with the CSM, and until after the LM jettison burn. After that, passive support was shifted to the CSM during TEC. Coverage was supplemented by the MARS 210-foot antenna which supplied TV and telemetry reception from the CSM during TLC and the lunar phase, and from the LM during the lunar phase including EVA.

Figure 3-37. GDS/GDSX Tracking Station

An operator error prevented correct time correlation tests at 56:27:00 GET (3.1.1.2).

An unscheduled handover to HAW was performed at 80:41:20 GET due to the loss of command capability (3.1.1.3).

A 48-minute loss of CSM downlink occurred during TEC due to a procedural error at MCC (3.1.1.4).

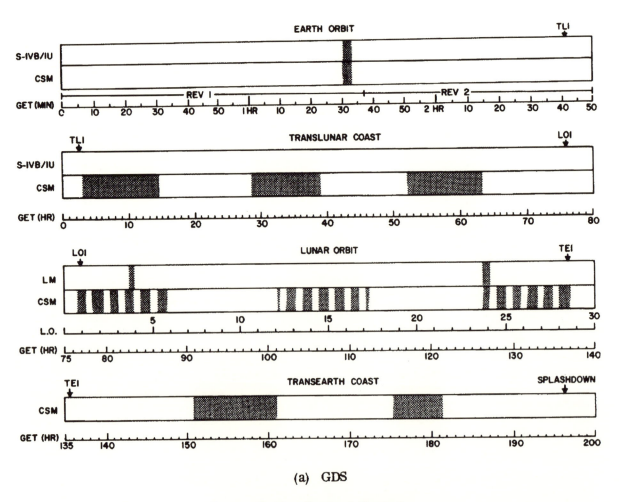

Figure 3-38. GDS/GDSX Mission Support

(a) GDS

Figure 3-39. Support Periods

3-48

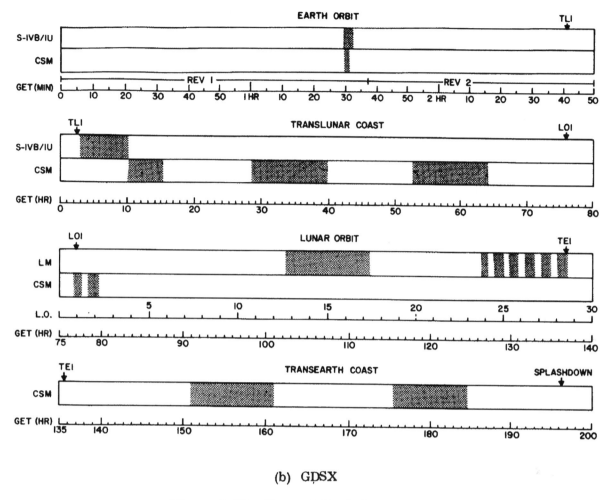

(b) GDSX

Figure 3-39. Support Periods (cont)

3.10.2 TRACKING

3.10.2.1 Goldstone (GDS) USB (Figure 3-40). During EO revolution 1, a TV transmission from the spacecraft was not obtained by GDS because of a defective patch cord in SDDS No. 1.

At acquisition following TLI (02:55:00 GET), downlink voice remoted to MCC was distorted and was subsequently bypassed. Investigation revealed that the input cable to voice demod No. 1 was defective and decreased the input level by several volts. Additionally, a substandard PM voice amplifier was discovered in voice demod No. 3.

During a GDS-to-MAD HGA handover at 04:02:00 GET, the GDS exciter operator used omni antenna handover procedures. Since HGA handover procedures were not used, it was necessary for MAD to initiate uplink sweep to acquire the spacecraft downlink signal. Later during the same view period, while the station was passively supporting the CSM, the low- and high-speed data from the TDP was flagged "bad" because of an incorrectly positioned data switch. Data between 04:40:06 and 06:23:30 GET was invalidated due to the operator error.

On the station's second view of TLC, 1 hour and 24 minutes of TDP backup tape was lost when the tape punch failed at 31:35:00 GET and the problem was not detected by the operator.

3-49

At 84:44:00 GET during LO 5, failure of a 24-volt dc power supply (acquisition bias) caused receiver data dropouts and eventual receiver No. 1 loss of signal. The exciter operator initiated two-way reacquisition while receiver No. 2 was in lock, sweeping the exciter before shorting the synthesizer loop. This caused the spacecraft HGA to slew off and the reacquisition procedure had to be reinitiated, resulting in delayed AOS.

At 127:57:40 GET, acquisition of the downlink signal for LO 27 was delayed for 6 minutes and 28 seconds when the exciter operator did not initiate the contingency procedure according to the NOD. While attempting two-way lock, loss of signal occurred when the receiver operator selected narrow-loop bandwidth prior to turning off the exciter sweep bias. Reacquisition of the downlink signal was achieved using wide bandwidth.

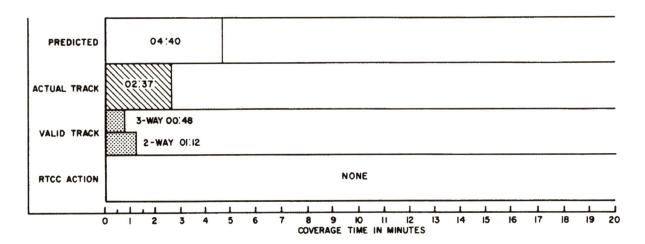

(1) EO -- CSM

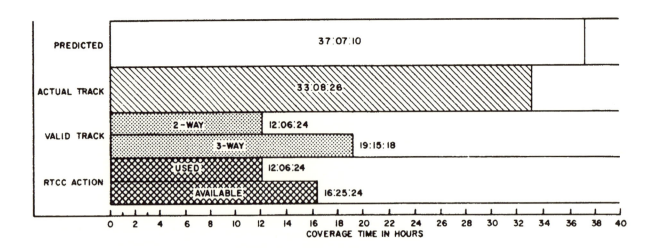

(2) TLC -- CSM

(a) System No. 1

Figure 3-40. GDS USB Tracking Coverage

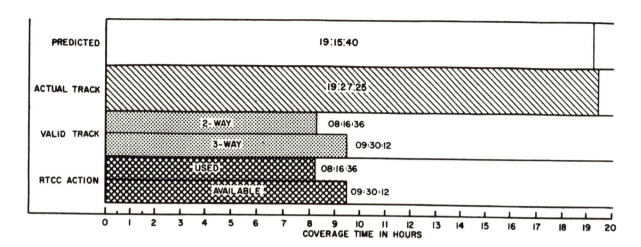

(3) Lunar Phase -- CSM

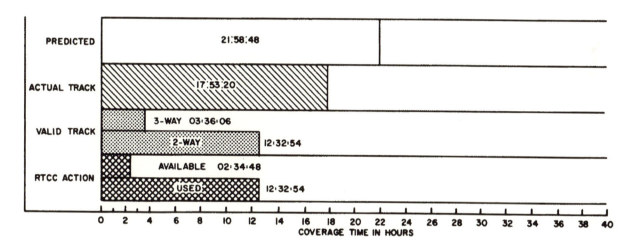

(4) TEC -- CSM

(a) System No. 1 (cont)

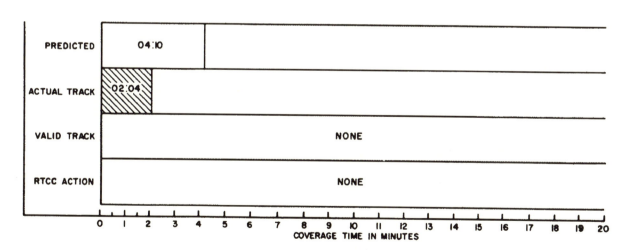

(1) EO -- IU

(b) System No. 2

Figure 3-40. GDS USB Tracking Coverage (cont)

3-51

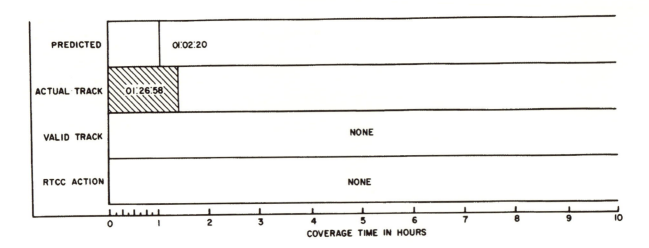

(2) Lunar Phase -- LM

(b) System No. 2 (cont)

Figure 3-40. GDS USB Tracking Coverage (cont)

3.10.2.2 <u>Goldstone Wing (GDSX) USB (Figure 3-41)</u>. Horizon masking on EO revolution 1 prevented GDSX from providing support of a TV transmission.

At 04:53:00 GET while actively supporting the IU, tracking data was invalidated and was rejected when the data was erroneously flagged as "LM".

At 05:50:00 GET, PA No. 1 faulted when the operator was increasing mission power from 2 to 10 kW. The fault was caused by overshooting the beam voltage meter relay setting while increasing power. The PA was reset and the beam voltage returned to the proper value. Momentary activation of the beam and body overcurrent fault interlocks caused this PA to be intermittently unreliable. PA No. 2 was brought up for prime command requirements throughout the remainder of the mission. Later, at 35:29:50 GET, PA No. 2 was momentarily disabled because of body overcurrent, causing loss of uplink capability.

Recycling the PA restored normal operation and tracking was resumed. Numerous PA failures were encountered throughout the MSFN during the mission (3.1.2.2).

While still in TLC at 64:13:00 GET, maser No. 2 became inoperative and was down for 3 hours. Cause of the failure was cryogenics contamination and it was necessary to replace the defective unit with a spare pre-cooled maser.

At 76:18:00 GET, after the start of the lunar phase, the high-speed servo tachometer was operating marginally. This unit was replaced postmission because only one spare was available and the high-speed mode was not required. After LM touchdown (103:18:40 GET), a hydraulic failure was encountered by GDSX. Antenna brakes were applied immediately while the antenna was still on track, thus preventing any data loss. When repairs to the hydraulics were completed, autotrack was resumed. The hydraulic difficulty was attributed to moisture in the emergency pushbutton safety interlocks and the problem was corrected when these emergency circuits were by-passed.

3-52

A handover to HSK was scheduled during the lunar stay while the LM was downlinking FM on the HGA. There was no procedure documented for this handover; a temporary procedure was developed that proved satisfactory. HSK used a small amount of uplink sweep while GDSX used the normal high-gain handover method. GDSX terminated uplink when HSK uplink lock was verified. A procedure for this handover is being developed and will be included in the MOM.

3.10.2.3 Radar. There is no C-band radar capability at GDS or GDSX.

3.10.3 TELEMETRY (Figure 3-42)

3.10.3.1 USB. Keyhole and terrain restrictions, low-elevation passes, fluctuating signal levels due to spacecraft antenna switching, and the PTC mode of flight caused some loss of telemetry data (3.1.3.1).

On launch day, decom No. 3 was unable to pass the telemetry CADFISS test in the terminal count. An invalid IU prime frame synchronization data pattern was being input to the 642B telemetry processor. A PCM error analysis incorporated within the telemetry operational program was initiated to isolate this problem. Two bent connector pins on a PC card in location 19C in the output data buffer were located and repaired. The bent pins were causing bits 9 and 10 of the prime frame synchronization

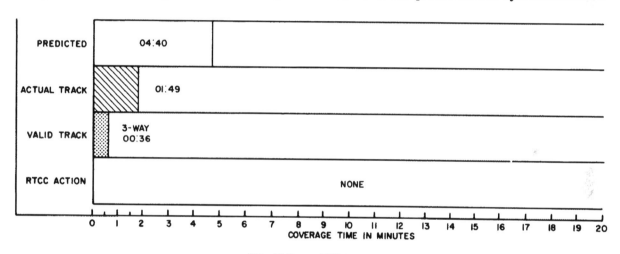

(1) EO -- CSM

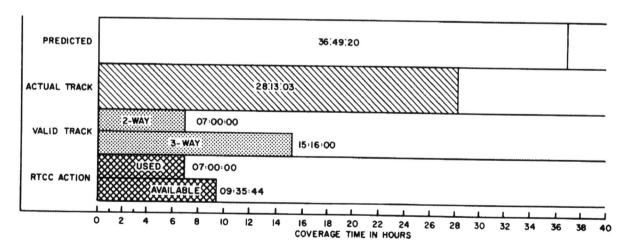

(2) TLC -- CSM

(a) System No. 1

Figure 3-41. GDSX USB Tracking Coverage

3-53

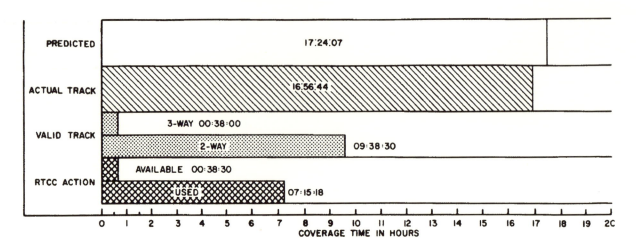

(3) Lunar Phase -- LM

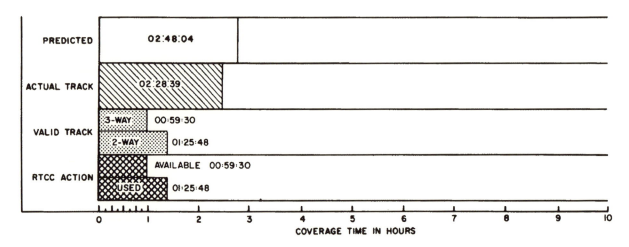

(4) Lunar Phase -- CSM

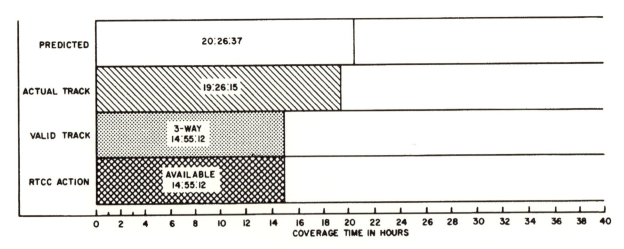

(5) TEC -- CSM

(a) System No. 1 (cont)

Figure 3-41. GDSX USB Tracking Coverage (cont)

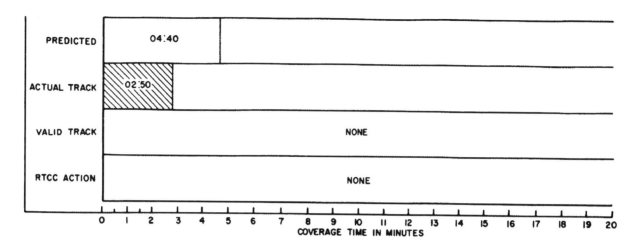

(1) EO -- IU

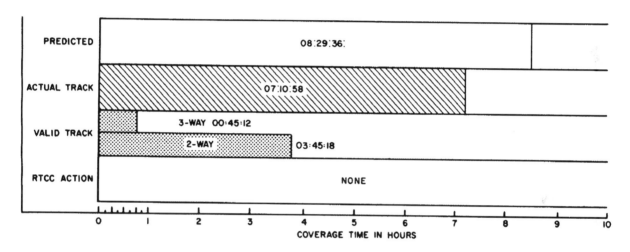

(2) TLC -- IU

(b) System No. 2

Figure 3-41. GDSX USB Tracking Coverage (cont)

pattern to be constantly reset at the telemetry computer input. The system was operational prior to liftoff with no further difficulty being experienced.

Two PCM operator errors did occur during support of the CSM prior to the LM descent and landing, but caused no mission impact. Five DSE dump messages were initiated listing central timing equipment errors which were erroneous by 48 hours. The errors were noticed by TIC at MCC who resolved the problem by voice contact with GDS. Also during this period, a LM PCM parameter readout was given to the TIC in error as an octal 001. The error was noted by other network stations and the correction (octal 175) was passed to the TIC. The operator confirmed the correction and acknowledged his error following another playback of the data on station.

MCC reported loss of subframe synchronization at 135:46:00 GET on all telemetry data received from GDS. The M&O reported that the time search routine was being used at the time and it caused a "no frame count" condition to exist with the telemetry computer. No impact resulted as MCC was using GWM CSM data as the prime source at

3-55

this time. The cause of this malfunction has not been determined though it is suspected to be caused by the time search routine being used while processing real-time data.

3.10.3.2 **VHF.** There is no VHF telemetry capability at GDS or GDSX.

3.10.3.3 **Telemetry Computer.** A telemetry computer fault occurred at 35:59:02 GET during TLC, but an automatic recovery was achieved without difficulty. No further difficulty was encountered until TEC when, at 176:26:00 GET, a fault occurred and automatic recovery was not achieved. This necessitated a manual recovery and resulted in a 3-second data loss. Another fault occurred at 176:29:00 GET while using the LUK routine (call up and print out for inspection any location in memory core) in testing for the previous fault. Three seconds of data was lost before handing over to MAD. The cause of these faults was not determined. Several telemetry computer faults were encountered by the MSFN throughout the mission (3.1.3.4).

A telemetry HSD problem occurred at 135:46:00 GET during the lunar phase. Both data lines seemed to lock up and continuously output one data frame while GDS was exercising a time search routine. The third time this routine was initiated, the program seemed to loop and remain in this condition. After loss of signal of the CSM dump data, the program automatically recovered. The time code translator has been observed to generate continuous external interrupts when switched from playback to search. Such a condition could cause a program loop and the repeated data as previously described. This problem is currently being investigated.

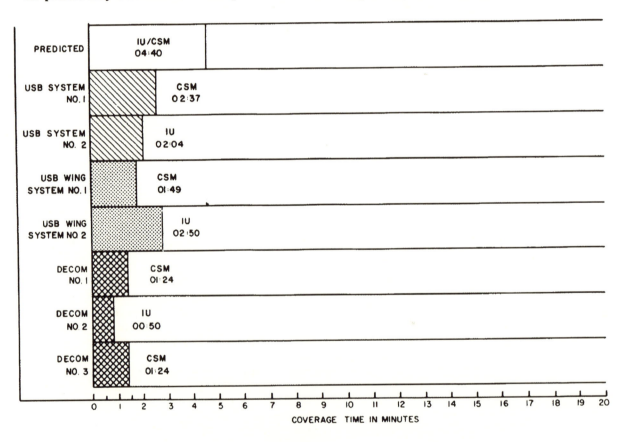

(a) EO

Figure 3-42. GDS/GDSX Telemetry Coverage

3-56

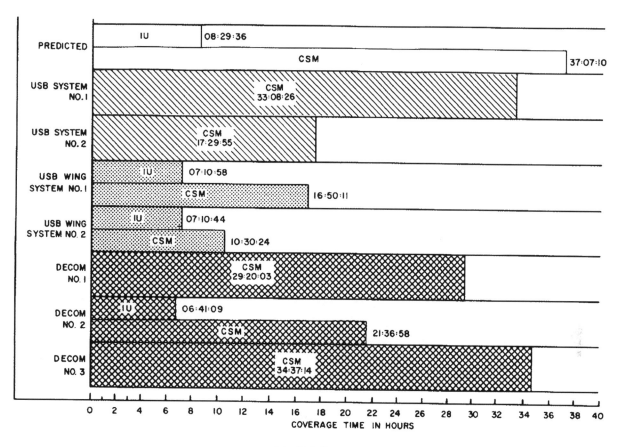

(b) TLC

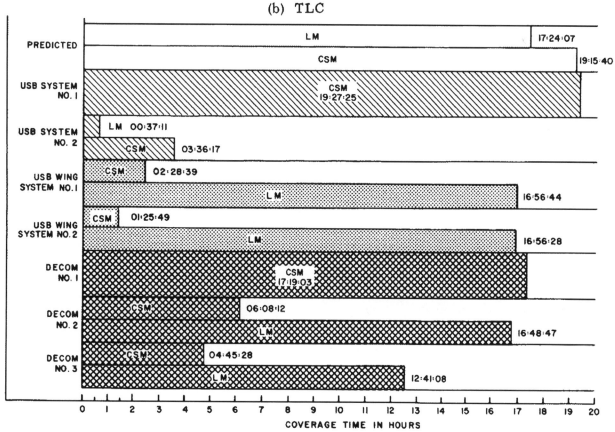

(c) Lunar Phase

Figure 3-42. GDS/GDSX Telemetry Coverage (cont)

3-57

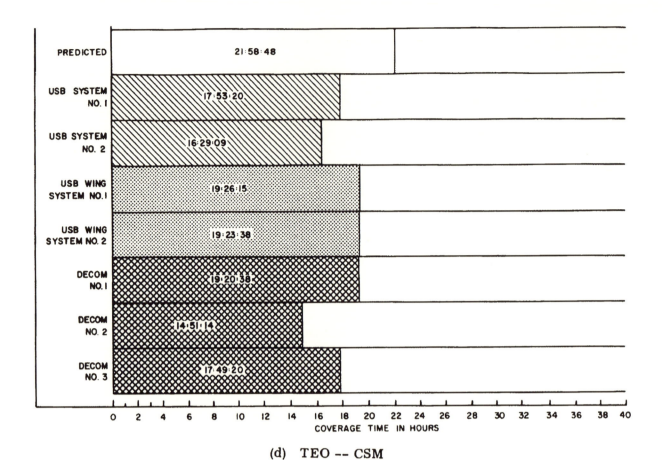

(d) TEO -- CSM

Figure 3-42. GDS/GDSX Telemetry Coverage (cont)

3.10.4 COMMAND

The only problem with the command system was the failure of a diode in the command verification circuit of the primary data multiplexer cabinet, which caused a loss of command capability and necessitated a handover to HAW at 80:41:20 GET (3.1.1.3). The first indication of a problem occurred at 80:35:22 GET when a bad UDB status printout was noted. A DSE tape stop command was sent three times with three ground rejects received. This command was verified on the fourth attempt. Three RTC's for HBR were uplinked with three ground rejects occurring. A command was then sent three times for "FM ON," and three ground rejects were again received. The command for HBR was again sent and verified. The "FM ON" command was again sent three times with three ground rejects. It was at this time that GDS handed over to HAW. Command capability to the spacecraft was lost until the handover was successfully completed. During this activity, the GDS command computer faulted at 80:36:00 GET, but automatic recovery was successful. Several computer faults were encountered by MSFN stations during the mission (3.1.4.1).

3.10.5 AIR-TO-GROUND COMMUNICATIONS

3.10.5.1 USB. Early in TLC at 03:00:00 GET, GDS failed to remote downlink voice to MCC for the first 10 minutes of the view period. The problem was a misinterpretation of NOD instructions pertaining to earth orbit procedure after the spacecraft had initiated TLI. The NOD instructions in question have been revised and clarified for subsequent missions.

3-58

Remoted voice was distorted until a handover to MAD at 04:00:00 GET. The distortion was caused by a defective cable to demod No. 1. GDS has revised the data demod test (SST-417G) to utilize both voice and tone as input sources. This change will enable earlier detection of such difficulty. Action has also been initiated to have this revised test implemented at all MSFN stations.

During the first view of TLC, several uplinks were not completed. Subsequent checks with NASCOM and MCC ComTech revealed that no problem existed at GDS. Also at this time, varying downlink voice remoted to MCC was received at levels from unreadable to loud and clear. The signal as monitored at GDS did not indicate this variation. It was determined that the problem occurred in an area between GSFC and MCC (Appendix A).

During LO 4, LM voice was selected by the ComTech in lieu of the required CSM voice. This operator error resulted in a 30-second loss of downlink voice capability. No mission impact resulted from this error because voice communication was not attempted at this time.

During EVA, an echo was evident on GOSS conference when CapCom was uplinking to the LM. This problem occurs only when both LM astronauts are on the extravehicular communications system. This situation was also observed at HSK and is presently under investigation by MSC (3.1.5).

A malfunction of the Quindar tone transmitter occurred during TEC (127:30:00 GET), causing improper operation in the PM lock indicator at GDSX. Adjustment of the Quindar resolved the problem with no mission impact.

On numerous occasions the MCC ComTech had to query GDS to determine the cause of distortion on GOSS conference. The problem existed at several MSFN stations; however GDS and MAD were the only stations to report that the problem was suspected of being caused by the VOGAA (3.1.5.2). The VOGAA is suspected of having caused distortion at several stations other than GDS and MAD; the problem is being investigated by MFED.

3.10.5.2 VHF. There is no VHF voice capability at GDS or GDSX.

3.11 GRAND BAHAMA ISLAND (GBM) (Figures 3-43 through 3-47)

3.11.1 GENERAL (Figures 3-43 through 3-45)

GBM provided support during launch, EO revolution 2, and the early portion of TLC. Support terminated when the station was released on July 16 at 10:12:00 GET. S-band voice and command capabilities were inhibited when the uplink carrier was terminated 30 seconds early during launch due to an operator error (3.1.1.1). Additionally, it was discovered postpass that the data modulator-demod outputs had not been recorded. This problem, attributed to an operator error, resulted because the recorder tieline had not been properly patched at the data terminal patch panel.

The VHF system and the command and telemetry computers experienced no problems.

3.11.2 TRACKING

3.11.2.1 USB (Figure 3-46). GBM tracked the CSM during launch, and the IU for all their remaining support. Due to an operator error, CSM tracking on launch was terminated 30 seconds early.

3.11.2.2 Radar. No requirements existed.

3-59

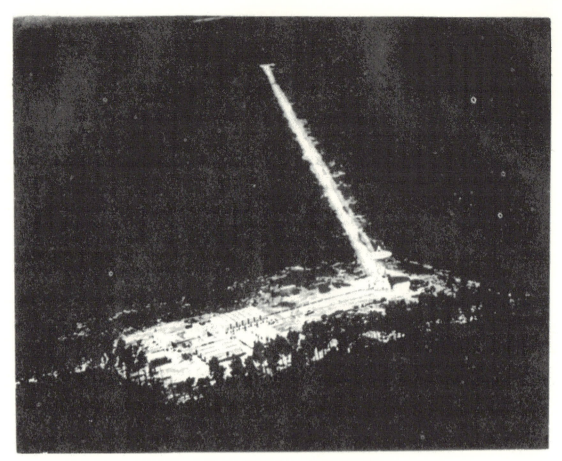

Figure 3-43. GBM Tracking Station

	EARTH ORBIT		TRANSLUNAR COAST				LUNAR PHASE																															TRA EA COA	
	REV		DAY				REVOLUTION																															DA	
SYSTEM	1	2	1	2	3	4	1	2	3	4	5	6	7	8	9	10	11	12	13	14	15	16	17	18	19	20	21	22	23	24	25	26	27	28	29	30	31	7	8
NO. 1	1	5	5																																				
AID NO. 1	5	5	6																																				
OM NO. 1	(5)	(5)																																					
OM NO. 2	1	(5)																																					
OM NO. 3	(6)	6																																					

1=CSM 5=IU 6=S-IVB
O=VHF TELEMETRY PROCESSED
☐=LM LUNAR STAY

Figure 3-44. GBM Mission Support

1.3 TELEMETRY (Figure 3-47)

1.3.1 USB. During launch, 20 seconds of S-band telemetry was lost when an
rator terminated the S-band carrier 30 seconds early. Reacquisition of the
nlink carrier was accomplished successfully after discovering the error (3.1.1

1.3.2 VHF

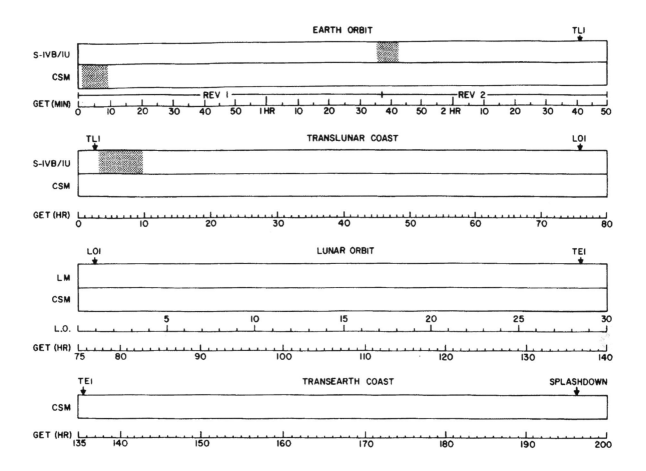

Figure 3-45. GBM Support Periods

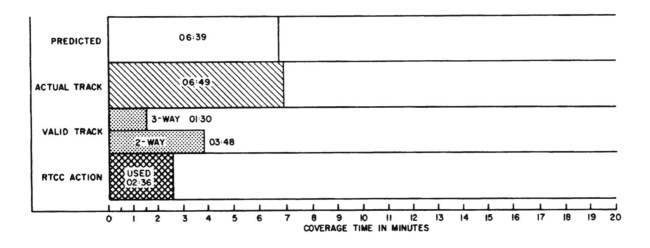

(a) EO -- IU

Figure 3-46. GBM USB Tracking Coverage

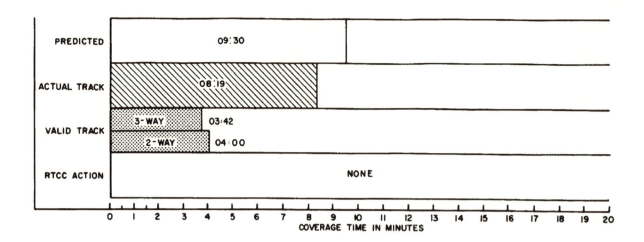

(b) EO -- CSM

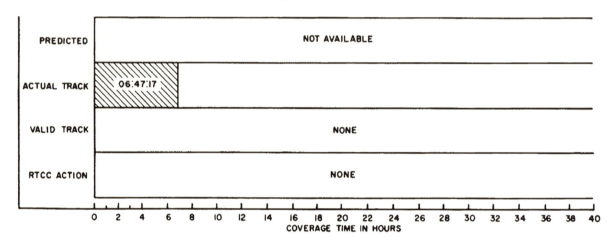

(c) TLC -- IU

Figure 3-46. GBM USB Tracking Coverage (cont)

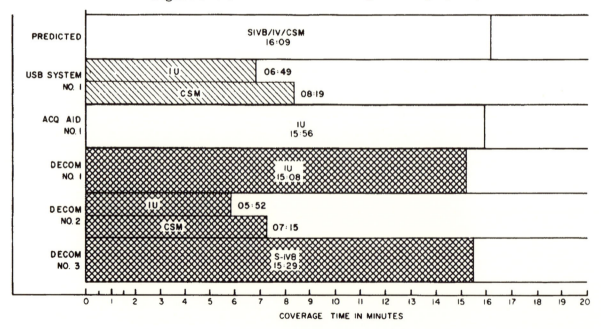

(a) EO

Figure 3-47. GBM Telemetry Coverage

3-62

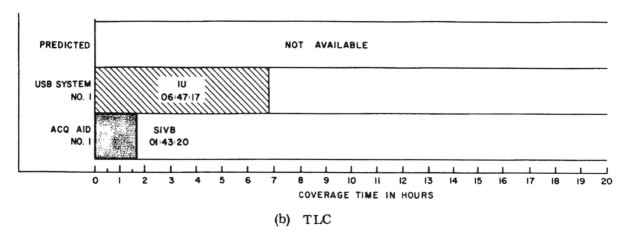

(b) TLC

Figure 3-47. GBM Telemetry Coverage (cont)

3.11.3.3 **Telemetry Computer**

3.11.4 COMMAND

USB command capability was lost for 30 seconds on launch when the uplink carrier was terminated early (3.1.1.1).

3.11.5 AIR-TO-GROUND COMMUNICATIONS

3.11.5.1 **USB.** No problems other than that mentioned under paragraph 3.11.1 were encountered.

3.11.5.2 **VHF**

3.12 **GUAM (GWM) (Figures 3-48 through 3-50)**

3.12.1 GENERAL (Figures 3-48 through 3-50)

GWM's support consisted primarily of backup station activities during TLC, the lunar phase, and TEC. The station did provide approximately 10 hours of two-way support of the CSM during TEC.

During TLC, broken pen bands on the servo recorder resulted in a loss of recorded data on two occasions. On launch day, approximately 1 hour of recorded data was interrupted because of broken pen bands for channels 5 (Y-acquisition error) and 6 (acquisition receiver AGC). Replacement of the pen bands restored the recording capability for these events. Two days later, the channel 5 pen band on the servo recorder broke again. Before replacement was accomplished, there was a 45-minute interruption to the Y-acquisition error recorded data. This problem occurred at several MSFN stations during the mission (3.1.1.5).

No problems in the areas of VHF telemetry, command, or A/G communications occurred at GWM.

3.12.2 TRACKING

3.12.2.1 **USB (Figure 3-51).** Early during TLC the system No. 1 range receiver became inoperative due to a clock lock relay failure at 18:58:00 GET. This failure necessitated the use of system No. 2 to support a time correlation test while the faulty relay was being replaced. The unit was operational at 19:32:00 GET.

3-63

Figure 3-48. GWM Tracking Station

Figure 3-49. GWM Mission Support

While supporting the CSM during LO-18, the 1218 computer faulted when the operator pressed the start switch for real-time track. The CADCPS program was reloaded and tracking resumed with no data loss. Another 1218 computer fault occurred during TEC while processing the 29-point acquisition message. Diagnostics were run and revealed no hardware malfunction. The program was reloaded and support was continued with no data loss. Several 1218 computer faults occurred throughout the MSFN during the mission (3.1.2.1).

While passively supporting the CSM during LO 20, the station PA shut down at 114:43:00 GET, inhibiting the uplink capability briefly. This shutdown was caused by heavy winds, accompanying a local rain storm, which initiated closure of the blower-operated heat exchange vane and tripped the interlock microswitch. The interlock switch was reset and the PA recycled to restore uplink capability. No mission impact resulted as GWM had no uplink requirement at this time. Later during this view period, a spark-gap arc in the PA high-voltage power supply caused a phase interlock shutdown to interrupt uplink capability at 191:02:06 GET. The PA was recycled and uplink capability restored. Several MSFN stations encountered PA problems during the mission (3.1.2.2).

3.12.2.2 **Radar.** GWM has no capability.

3-64

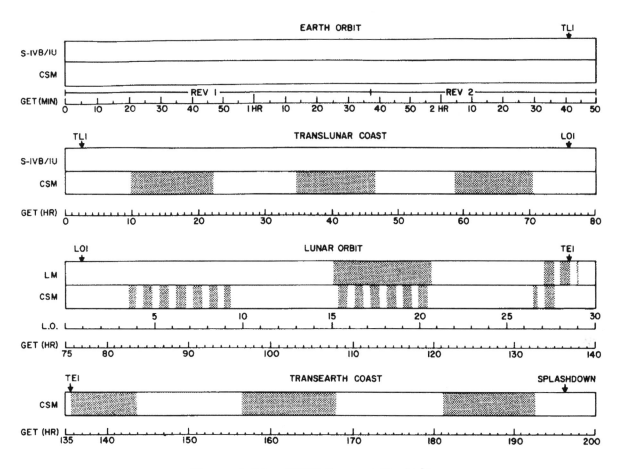

Figure 3-50. GWM Support Periods

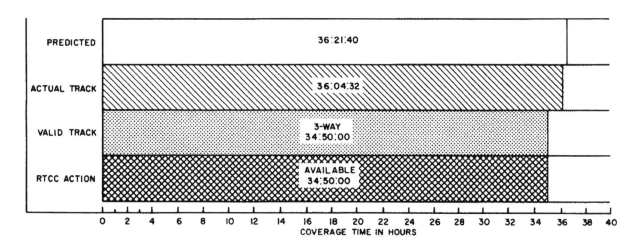

(1) TLC -- CSM

(a) System No. 1

Figure 3-51. GWM USB Tracking Coverage

3-65

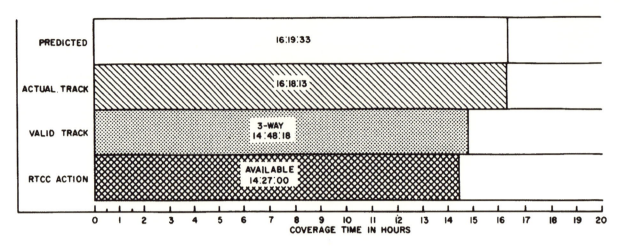

(2) Lunar Phase -- CSM

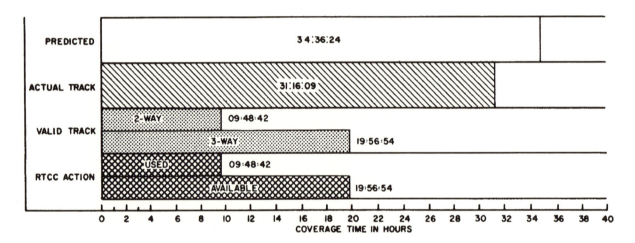

(3) TEC -- CSM

(a) System No. 1 (cont)

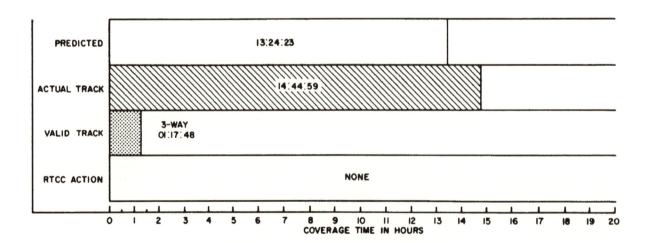

(b) System No. 2, Lunar Phase -- LM

Figure 3-51. GWM USB Tracking Coverage (cont)

3.12.3 TELEMETRY (Figure 3-52)

3.12.3.1 USB. As in previous Apollo missions, data dropouts were caused by weak and fluctuating signals as a result of the unfavorable aspect angle of the spacecraft antenna, the spacecraft utilizing the PTC mode of flight, use of the omni antenna at extreme distances, and low to high bit rate changes (3.1.3.1). Approximately 14 seconds of LM data was lost during lunar liftoff due to degraded received signal (3.1.3.2).

3.12.3.2 VHF

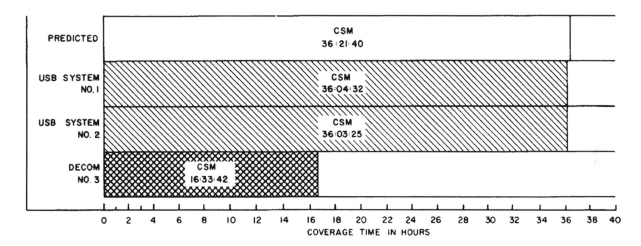

(a) TLC

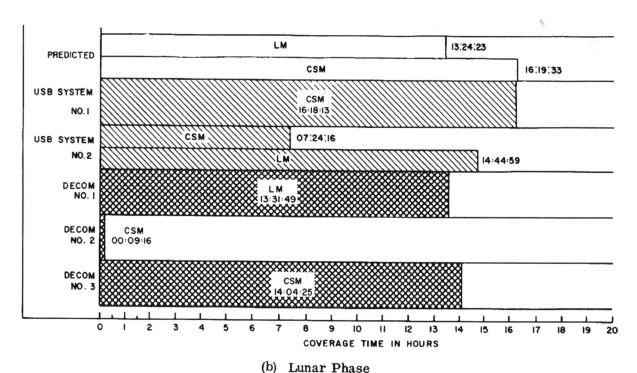

(b) Lunar Phase

Figure 3-52. GWM Telemetry Coverage

3-67

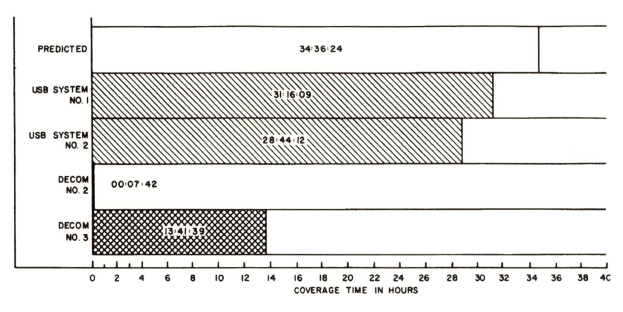

(c) TEC -- CSM

Figure 3-52. GWM Telemetry Coverage (cont)

3.12.3.3 Telemetry Computer. During TLC at approximately 16:00:00 GET, GWM was supporting real-time telemetry and experienced a telemetry computer fault which did not light the fault indicator (3.1.3.4). There was no mission impact as a result of the fault.

During the entire mission, the CADFISS test conductor was unable to remotely select the telemetry format for H-70 telemetry CADFISS testing (3.1.3.4).

3.12.4 COMMAND

3.12.5 AIR-TO-GROUND COMMUNICATIONS

3.13 GUAYMAS (GYM) (Figures 3-53 through 3-57)

3.13.1 GENERAL (Figures 3-53 through 3-55)

GYM provided support during EO revolution 1, TLC, the lunar phase, and TEC. Throughout the mission, several pens on Brush recorder No. 8 of the systems monitor failed. The pens were promptly replaced and recording interruptions were kept to a minimum. This problem was also experienced by other stations (3.1.1.5).

An operator error at GDS prevented correct time correlation tests at 56:27:00 GET (3.1.1.2).

While passively tracking the CSM during LO 3, a loss of downlink signal was encountered due to an unscheduled handover from GDS to HAW (3.1.1.3).

A 48-minute loss of CSM downlink occurred during TEC due to a procedural error at MCC (3.1.1.4).

No problems were encountered by the telemetry computer or A/G communications during the mission.

3-68

Figure 3-53. GYM Tracking Station

Figure 3-54. GYM Mission Support

3.13.2 TRACKING

3.13.2.1 <u>USB (Figure 3-56)</u> Active S-band tracking of the IU was performed during EO revolution 1 and the first hour of TLC. Passive tracking of the IU continued until batteries on board the vehicle were depleted. After IU tracking terminated, the CSM was passively tracked for the remainder of TLC and during TEC. The CSM and LM were passively tracked during the lunar phase.

For EO revolution 1, the SCM was received 10 minutes prior to AOS instead of the 30 minutes required by the NOD. The SCM instructed GYM to track the IU with the acquisition receiver tuned to a CSM frequency. The 10 minutes did not allow the station sufficient time to reconfigure and calibrate equipment. Therefore, the reconfiguration was not attempted. Consequently, the S-band IU PCM signals were not recorded. However, no data was lost because the same parameters were recorded using VHF telemetry.

Vehicle identification during EO revolution 1 in the TDP was set to CSM instead of IU due to an operator error. This caused data being transmitted from the station to be

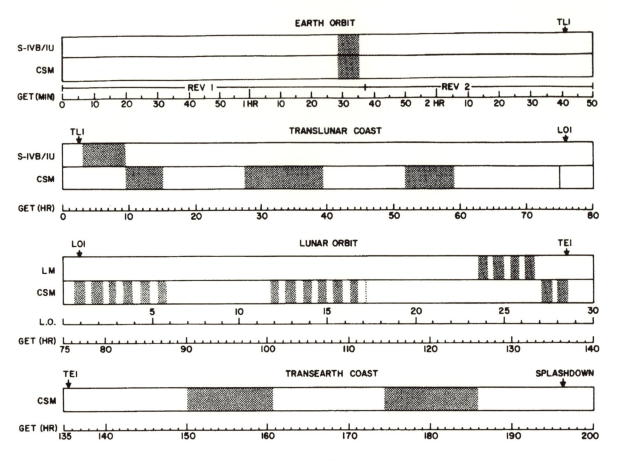

Figure 3-55. GYM Support Periods

incorrectly tagged. The error was corrected by repositioning the TDP vehicle identification switch. If the prepass checklist had been strictly adhered to, this operator error would not have occurred.

At the beginning of TLC at 02:23:00 GET, the sweep of receiver No. 2 was inoperative due to a dc isolation amplifier failure. The amplifier was replaced and the receiver restored. No data was lost because the station was providing passive support.

During TEC at 153:23:00 GET, relay K-11 of receiver No. 1 failed and indication of receiver lock was not available until the defective relay was replaced. This minor problem was corrected without causing loss of track.

3.13.2.2 Radar. GYM has no capability.

3.13.3 TELEMETRY (Figure 3-57).

3.13.3.1 USB. CSM S-band telemetry processing was hampered during the mission whenever the omni antenna on board the spacecraft was used, or the PTC mode of operation was initiated. In addition, CSM and LM S-band telemetry received signal strength decreased when bit rates changed from low to high. These problems caused the majority of data losses throughout the network. (3.1.3.1).

3.13.3.2 VHF. Received VHF telemetry signal strength decreased as the range to the S-IVB and IU increased during TLC. During the first view period following TLI (starting at 02:56:00 GET), VHF signals were very noisy because of a local electrical storm and it was difficult to maintain decom lock.

3-70

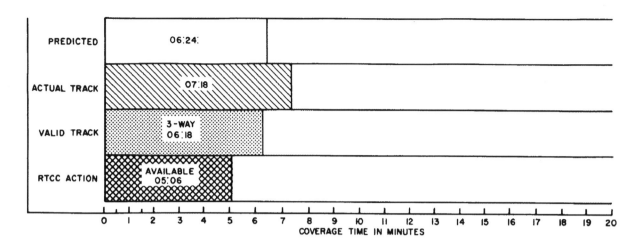

(1) EO -- IU

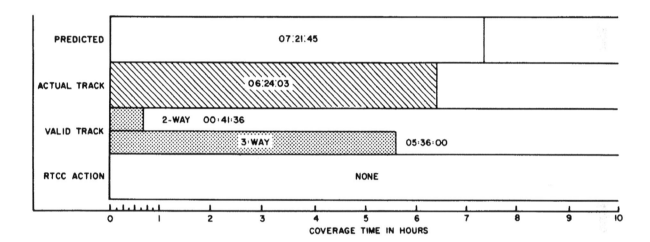

(2) TLC -- IU

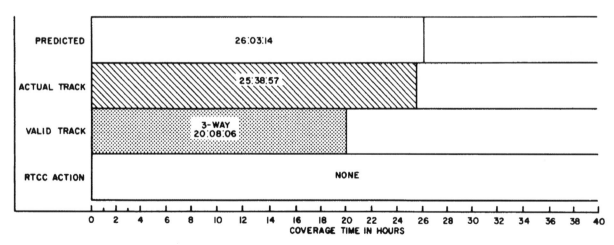

(3) TLC -- CSM

(a) System No. 1

Figure 3-56. GYM USB Tracking Coverage

3-71

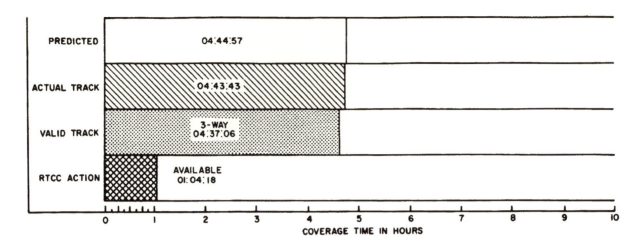

(4) Lunar Phase -- LM

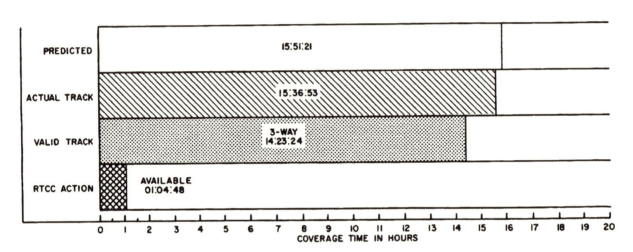

(5) Lunar Phase -- CSM

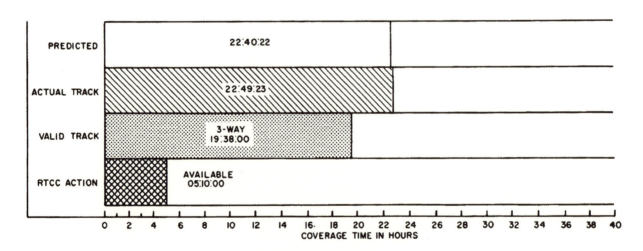

(6) TEC -- CSM

(a) System No. 1 (cont)

Figure 3-56. GYM USB Tracking Coverage (cont)

3-72

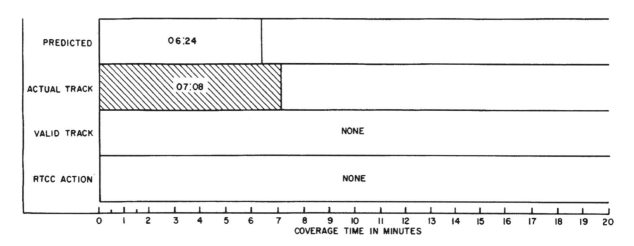

(b) System No. 2, EO -- CSM

Figure 3-56. GYM USB Tracking Coverage (cont)

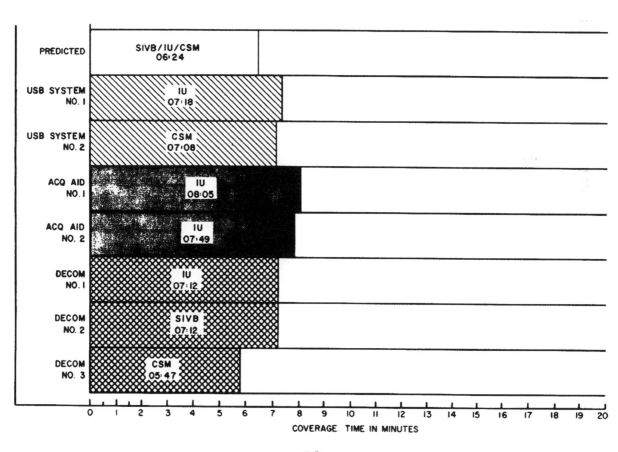

(a) EO

Figure 3-57. GYM Telemetry Coverage

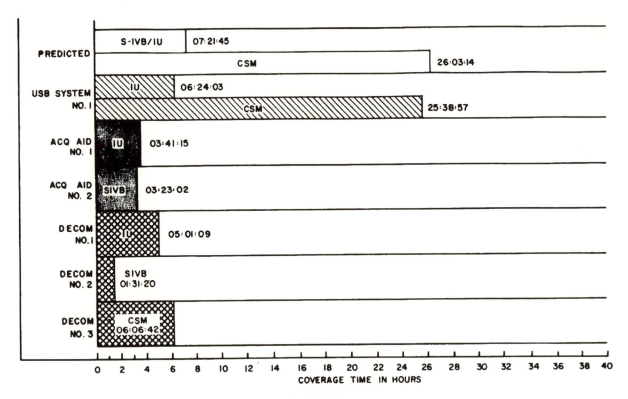

(b) TLC

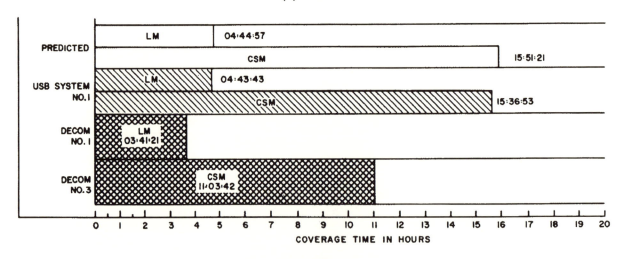

(c) Lunar Phase

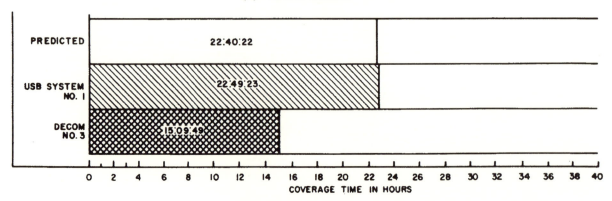

(d) TEC -- CSM

Figure 3-57. GYM Telemetry Coverage (cont)

3-74

3.13.3.3 Telemetry Computer

3.13.4 COMMAND

GYM did not provide any USB commanding during the mission. During the lunar phase, the command computer faulted. An automatic recovery was successful in restoring it to an operational status. Several other stations encountered command computer faults during the mission (3.1.4.1).

3.13.5 AIR-TO-GROUND COMMUNICATIONS

3.14 HAWAII (HAW) (Figures 3-58 through 3-63)

3.14.1 GENERAL (Figures 3-58 through 3-60)

HAW was called on suddenly to provide support of the critical S-IVB TLI burn when the MER was unable to fulfill its primary mission requirements due to equipment malfunctions, and again during LO 3 in support of the CSM due to a UDB failure at GDS. When the contingency handover from GDS was initiated for LO 3, the R/E operator at HAW erred by using omni antenna instead of HGA procedures. This caused the loss of tracking, command, voice, and telemetry capabilities for approximately 6 minutes (3.1.1.3).

An operator error at GDS prevented correct time correlation tests at 56:27:00 GET (3.1.1.2).

At 78:58:00 GET during LO 2, the pen for channel No. 3 of the Brush recorder failed. The AGC events recording was interrupted and the pen was replaced. A similar problem occurred at several MSFN stations (3.1.1.5).

Figure 3-58. HAW Tracking Station

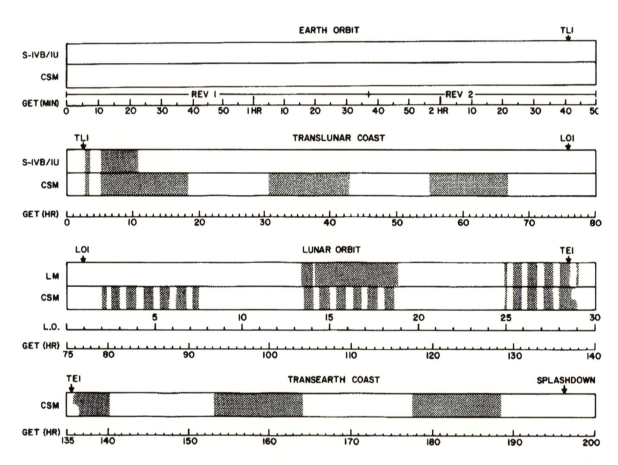

Figure 3-59. HAW Mission Support

Figure 3-60. HAW Support Periods

3.14.2 TRACKING

3.14.2.1 USB (Figure 3-61). While performing the SRT during TLC at 27:31:00 GET, the Mark 1A ranging unit for system No. 2 would not advance beyond program state "1." The problem was caused by a bad circuit module in the program and acquisition circuits. The faulty module was replaced and the unit restored to an operational status prior to the start of tracking operations.

3-76

At 33:14:00 GET during TLC, an anomaly was observed while performing CADFISS testing. The TDP output consisted of erroneous system No. 2 destruct Doppler data. This problem was traced to a "times 10" multiplier board failure in the time interval counter. The defective board was replaced upon receipt of a substitute at 93:06:00 GET. For the interim period, HAW was unable to support CADFISS boresight tests in the dual mode, but was able to operate in the single mode and test each system individually.

3.14.2.2 Radar (Figure 3-62). The FPS-16 tracked the IU during the early portion of TLC. The radar acquired the IU after the completion of the TLI burn. There were no signal dropouts, but lobing signals were encountered throughout the pass. During a second view period on the first day of TLC, only weak and sporadic returns were observed from the IU beacon and the radar was unable to acquire.

3.14.3 TELEMETRY (Figure 3-63)

3.14.3.1 USB. Signal losses were encountered while supporting the CSM during the coast and lunar phases whenever the omni antenna on board the spacecraft was used to transmit telemetry data. Nominal changes in received signal strength were noted when HBR was used by the CSM for data transmissions. The latter applied equally to the

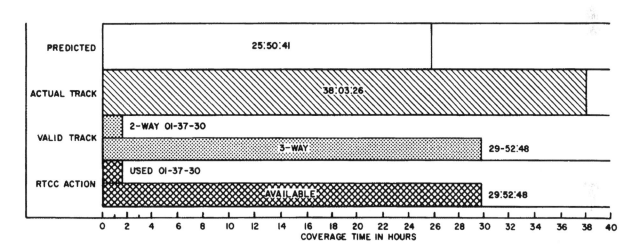

(1) TLC -- CSM

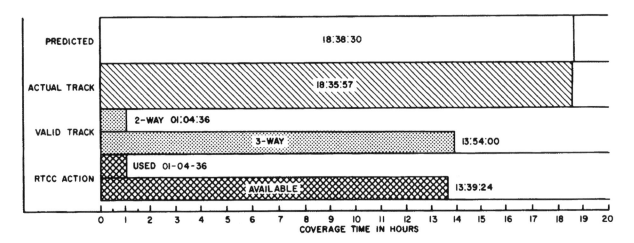

(2) Lunar Phase -- CSM

(a) System No. 1

Figure 3-61. HAW USB Tracking Coverage

3-77

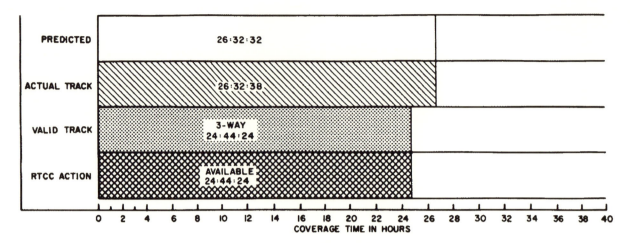

(3) TEC -- CSM

(a) System No. 1 (cont)

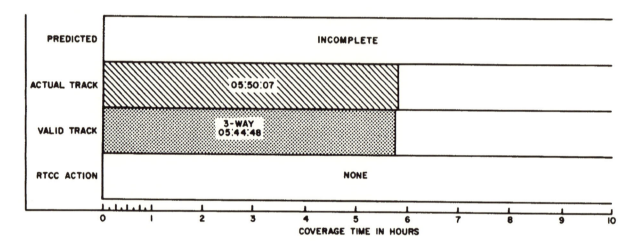

(1) TLC -- IU

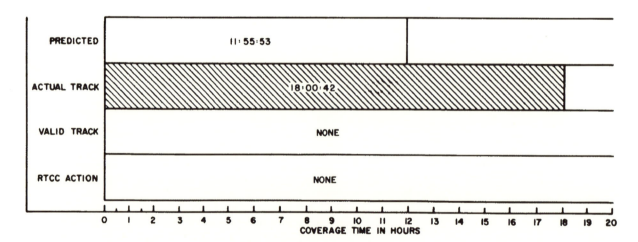

(2) Lunar Phase -- LM

(b) System No. 2

3-61. HAW USB Tracking Coverage (cont)

3-78

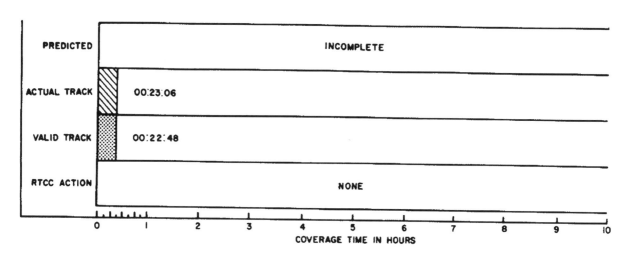

Figure 3-62. HAW FPS-16 Tracking Coverage, TLC -- IU

LM received signal strength during its lunar activities. Also, during the coast phases, whenever PTC operations were initiated by the CSM, degradation of received signals was noted (3.1.3.1).

3.14.3.2 **VHF**. Acq aids No. 1 (TELTRAC) and No. 2 (AGAVE retrofit) provided IU and S-IVB VHF telemetry support, respectively, during the last minute of TLI and approximately the first 25 minutes of TLC. A second view period, supported in the same configuration, began after the CSM evasive maneuver and continued until HAW VHF telemetry requirements were terminated at 06:28:00 GET.

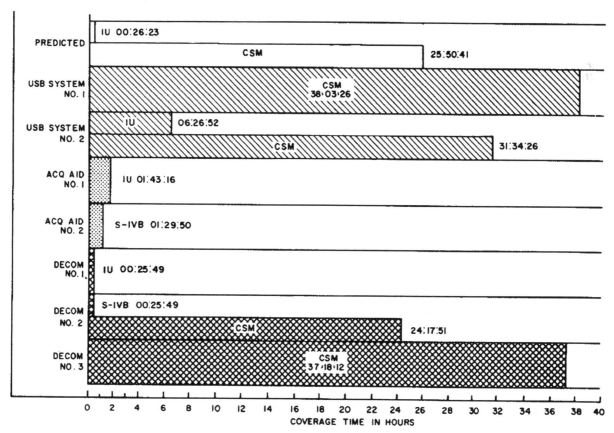

(a) TLC

Figure 3-63. HAW Telemetry Coverage

3-79

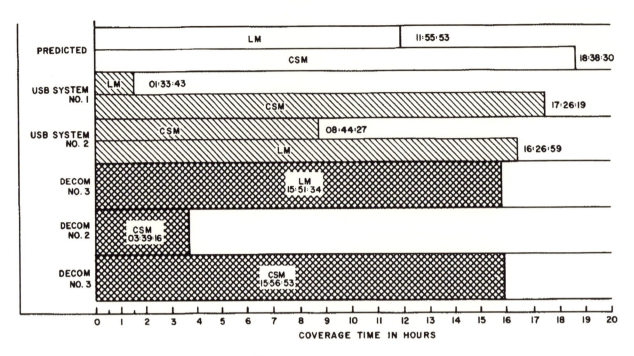

(b) Lunar Phase

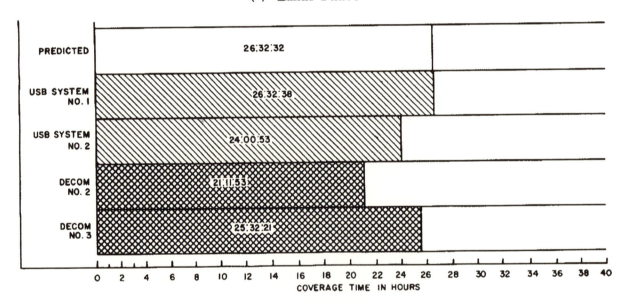

(c) TEC -- CSM

3-63. HAW Telemetry Coverage (cont)

3.14.3.3 <u>Telemetry Computer</u>. Near the beginning of TLC, HAW experienced a telemetry computer fault which did not illuminate the fault light. A similar problem was experienced at GWM (3.1.3.4).

During the entire mission, the CADFISS test conductor was unable to remotely select the telemetry formats for H-70 telemetry CADFISS testing at HAW, GWM, and MIL (3.1.3.4).

3.14.4 COMMAND

.14.5 AIR-TO-GROUND COMMUNICATIONS

ne station was in a remote configuration during TLC and the lunar phase. Only the
)link was remoted during the lunar phase.

.15 HONEYSUCKLE CREEK/HONEYSUCKLE CREEK WING (HSK/HSKX)
(Figures 3-64 through 3-69)

.15.1 GENERAL (Figures 3-64 through 3-66)

SK provided active support of the CSM during EO revolution 1, TLC, the lunar phase
nd TEC. Passive support for this vehicle was provided for TLC, the lunar phase, and
EC. In addition, both active and passive support was provided for the LM during its
inar stay.

uring TEC at 168:15:01 GET, the HSK exciter bias was decayed in wide bandwidth
y operator error while the receiver was in narrow-loop bandwidth, causing HGA
andover loss of signal. Reacquisition was accomplished promptly.

SKX supported the CSM actively and passively during the lunar phase and TEC, and
assively for TLC. The LM was actively supported after the TEI burn until the
:acking requirement for this vehicle was terminated. Limited passive support was
rovided for the LM during the lunar phase.

Figure 3-64. HSK/HSKX Tracking Stations

There were no problems encountered with the telemetry computer during support activity.

At 83:30:45 GET, recorder No. 4 at HSKX could not be operated in the "times 0.01" position because of a broken wire. The recorder was run at a speed of 5 mm/second until repairs were completed. Numerous systems monitor problems were encountered throughout the MSFN (3.1.1.5).

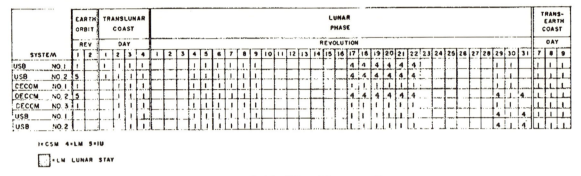

Figure 3-65. HSK/HSKX Mission Support

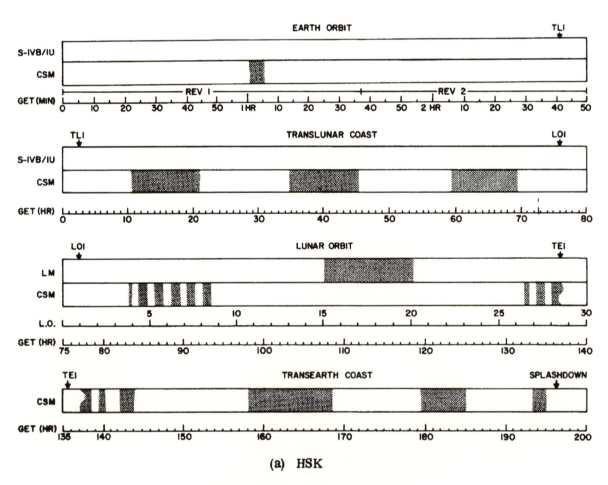

(a) HSK

Figure 3-66. Support Periods

3-82

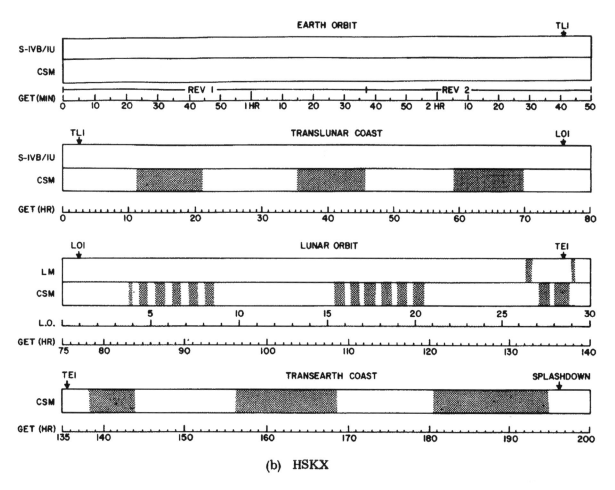

(b) HSKX

Figure 3-66. Support Periods (cont)

3.15.2 TRACKING

3.15.2.1 Honeysuckle Creek (HSK) USB (Figure 3-67). During TLC at 20:50:00 GET, tracking was momentarily interrupted when the antenna slewed off in the X-axis while autotracking. The antenna was returned to the proper position after program track was initiated. Postpass investigation revealed a corroded connector on the X-axis channel cable. As a temporary solution the connector was cleaned and the problem did not recur. After a postmission investigation, however, the connector was replaced. Near the end of the lunar phase (132:15:00 GET), a 10-dB discrepancy in cooled paramp gain was noted, apparently caused by a decrease in klystron pump power. Downlink signal degradation resulted, so the paramp was retuned. However, at the beginning of TEC, HSK noted an increase in paramp system noise, which caused a loss of approximately 5 minutes of CSM HBR data. Two-way lock was maintained, although paramp gain dropped sharply and oscillations were observed. Contingency handover between HSK and HSKX was initiated and accomplished in order for HSK to reconfigure from cooled to uncooled paramp operation. Over an hour was required to complete the changeover, after which tracking was returned to HSK (3.1.2.3).

3.15.2.2 Honeysuckle Creek Wing (HSKX) USB (Figure 3-68). Interruption of timing signals of two stripchart recorders occurred at 09:58:00 GET. The problem was a faulty timing cable. Repair to this cable restored normal timing annotation to the recorders at 16:40:00 GET.

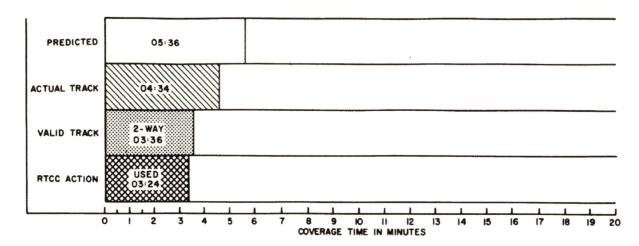

(1) EO -- CSM

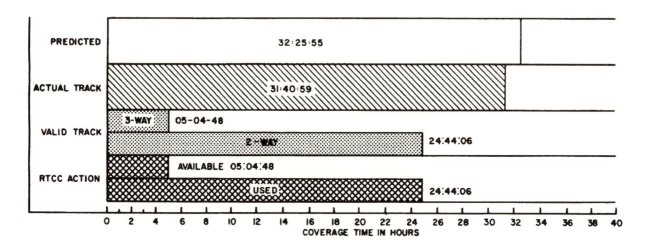

(2) TLC -- CSM

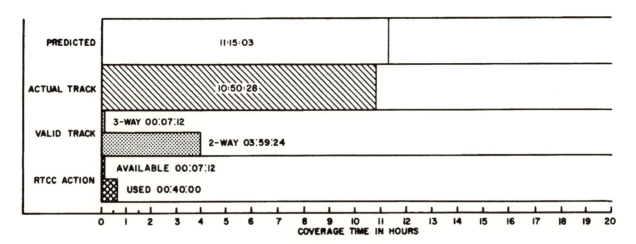

(3) Lunar Phase -- LM

(a) System No. 1

Figure 3-67. HSK USB Tracking Coverage

3-84

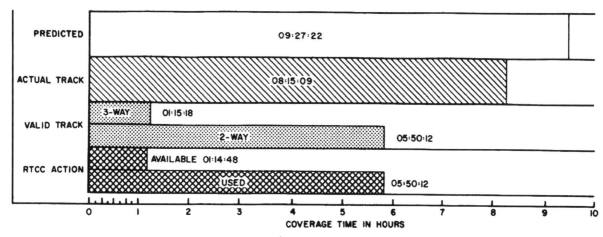

(4) Lunar Phase -- CSM

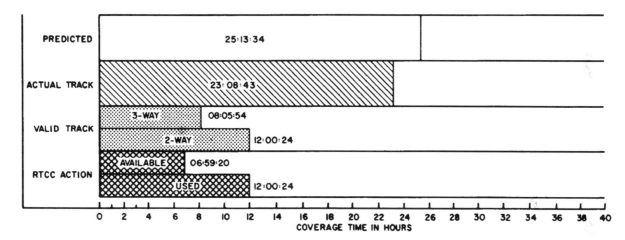

(5) TEC -- CSM

(a) System No. 1 (Cont)

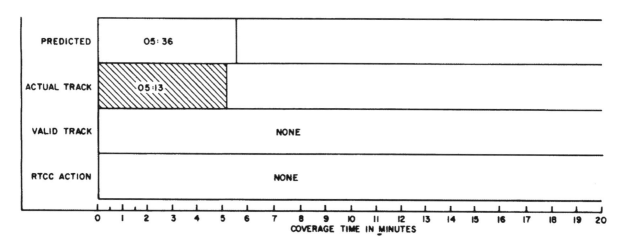

(b) System No. 2, EO -- IU

Figure 3-67. HSK USB Tracking Coverage (cont)

3-85

At 42:52:00 GET, the supply control system of PA No. 2 was severely damaged from an electrical fire caused by a short circuit in the 400-volt, three-phase power supply. Considerable damage was sustained to internal wiring, along with some melting and fusing of the metal work and cabinet panels. Uplink capability was lost using this PA, but PA No. 1 was available and was brought into service for command and voice transmissions while repairs were being performed on the damaged unit. The DSN is responsible for supporting M&O requirements of HSKX; parts and technical assistance were provided by the DSN's Tidbinbilla station near Woomera. PA No. 2 was restored to an operational status in less than 38 hours by a commendable joint effort of HSKX and Tidbinbilla personnel (3.1.2.2).

During TEC at 138:58:00 GET, erratic range data from system No. 1 was noted. The USB was reconfigured and system No. 2 provided all remaining tracking support. The cause of the problem could not be determined.

3.15.2.3 <u>Radar</u>. No capability exists at either HSK or HSKX.

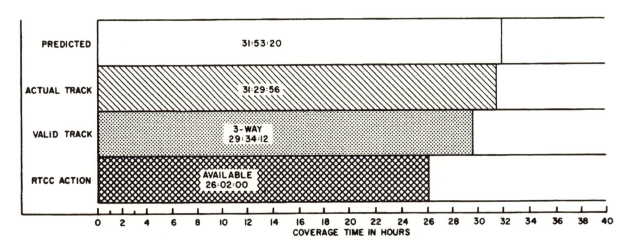

(1) TLC -- CSM

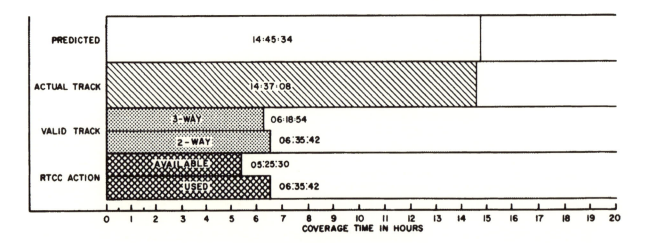

(2) Lunar Phase -- CSM

(a) System No. 1

Figure 3-68. HSKX USB Tracking Coverage

3-86

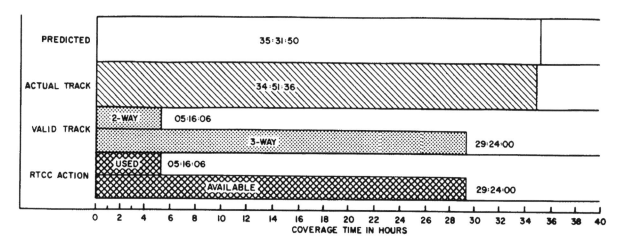

(3) TEC -- CSM

(a) System No. 1 (cont)

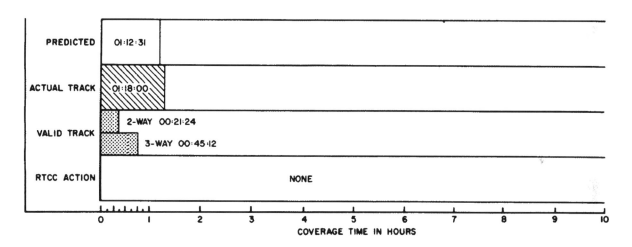

(b) System No. 2, Lunar Phase -- LM

Figure 3-68. HSKX USB Tracking Coverage (cont)

3.15.3 TELEMETRY (Figure 3-69).

3.15.3.1 USB. Real-time and dump data from the CSM were processed during EO, TLC, the lunar phase, and TEC. Real-time LM data was processed during the lunar phase, including the lunar stay, and until requirements for this vehicle were terminated. Limited support for EASEP was required during the final portion of TEC. Data loss other than that caused by spacecraft antenna usage was minimal (3.1.3.1). The primary cause of data loss was the HSK paramp problem already described in paragraph 3.15.2.1. Handover to the wing station minimized loss of telemetry data.

A unique problem was experienced during the premission phase when difficulty with the slow-scan TV converter was encountered. The problem was the inability to maintain the synchronized lock loop during test on July 7. On July 10, after realigning all circuitry and tightening all loose connections in the converter, the unit was declared "red, can support." The disc recorder in this converter became intermittent on

3-87

July 11 and failed completely on July 14. In the meantime, an emergency request went out to the network for a PC card required for this repair because a spare was not available at the depot. GDS shipped a card to the depot for trans-shipment, but it was lost enroute. GWM was requested to ship a spare, but a delay was experienced because of aircraft difficulty. The PC card finally arrived at HSK on July 17 and the scan-converter was declared "green" at 15:28:00 GET, early in TLC.

During TLC at 40:58:00 GET, 2 minutes of biomed data being remoted by HSK was lost due to loss of decom lock when a bit rate change from low to high occurred. This dropout resulted in biomed not being input to the IRIG VCO used for remoting data to MCC via the voice data line. The station was prime for biomed support and no backup was available.

The narrowband signal conditioner of decom No. 2 at HSK was reported to be operating in a marginal condition at 114:57:00 GET while supporting lunar surface activity of the LM. Fluctuating signals were being received which caused the decom to drop lock when signal strengths decreased below the threshold level of the signal conditioner. MCC elected to change the source of this LM real-time data to GWM and no further problems were indicated.

Approximately 14 seconds of LM data was lost during lunar liftoff due to degraded received signal (3.1.3.2).

3.15.3.2 VHF. HSK and HSKX have no VHF capability.

3.15.3.3 Telemetry Computer

3.15.4 COMMAND

The command computer faulted during TEC at 137:56:00 GET. The cause of the fault is unknown but recovery was achieved using CBARF (3.1.4.1). RTC histories were lost for 22 minutes, but the OUCH program recovered the histories postpass.

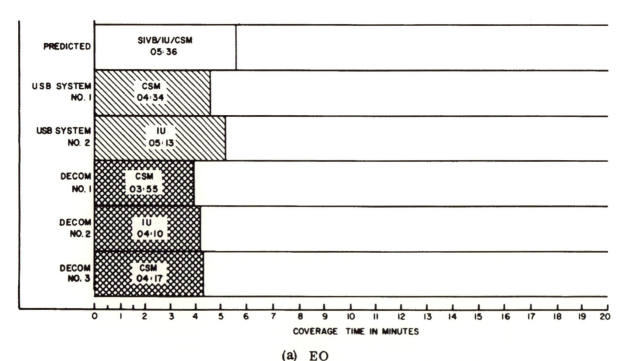

(a) EO

Figure 3-69. HSK/HSKX Telemetry Coverage

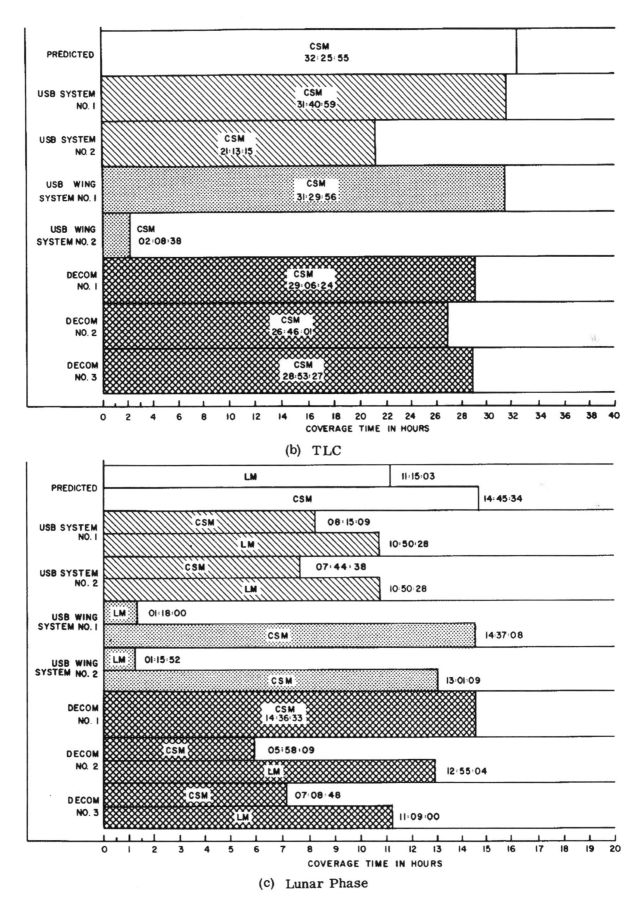

(b) TLC

(c) Lunar Phase

Figure 3-69. HSK/HSKX Telemetry Coverage (cont)

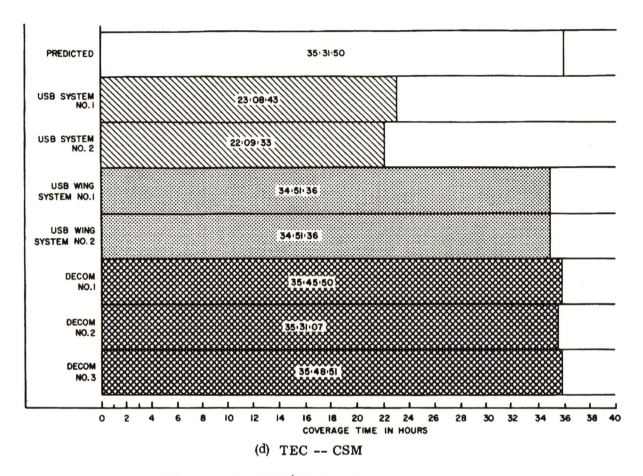

(d) TEC -- CSM

Figure 3-69. HSK/HSKX Telemetry Coverage (cont)

3.15.5 AIR-TO-GROUND COMMUNICATIONS

HSK provided S-band A/G communication support during all phases including the lunar stay.

During the lunar stay at 114:04:13 GET, cross-coupling was noted which caused an echo on the GOSS conference when the CapCom was uplinking to the LM. The only time this problem occurred was when both LM astronauts were on the extravehicular communications system; the problem is being investigated by MSC (3.1.5).

RTC was unable to command an omni antenna switch at 152:50:00 GET during TEC. This problem resulted in a 10-minute loss of voice from the spacecraft before successfully completing the antenna switch.

3.16 HUNTSVILLE (HTV) (Figures 3-70 through 3-75)

3.16.1 GENERAL (Figures 3-70 through 3-72)

The only mission requirement for HTV was to support the CM during reentry. There was no requirement for VHF support. No problems were experienced with A/G communications.

3-90

Figure 3-70. USNS Huntsville

Figure 3-71. HTV Mission Support

.16.2 TRACKING

.16.2.1 USB (Figure 3-73). Acquisition of the CM was achieved as the spacecraft emerged from S-band blackout by slaving to the CAPRI radar. Tracking was intermittent because of the supporting ARIA uplink carrier sweeping.

.16.2.2 Radar (Figure 3-74). During reentry, the CM was acquired within a few seconds of the predicted view time. Acquisition was achieved with the CAPRI slaved to the computer-driven IRV just after the CM broke the horizon. Almost immediately radar contact with the spacecraft was lost, but was quickly reacquired by slaving to the optical director. After reacquisition, 4 minutes and 8 seconds of solid track was accomplished, and preliminary indications reveal data quality to be well within acceptable limits.

.16.3 TELEMETRY (Figure 3-75)

.16.3.1 USB. Reentry support was the only requirement, and approximately 40 seconds of decom lock was obtained with 12 seconds of dropout reported. This substandard performance was attributed to the signal strength dropping below decom threshold.

3-91

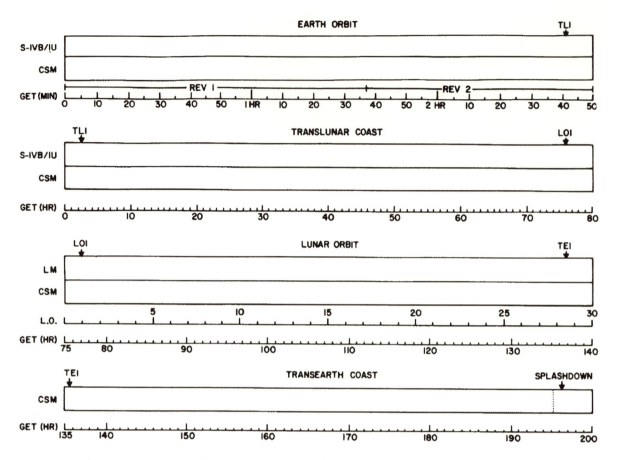

Figure 3-72. HTV Support Periods

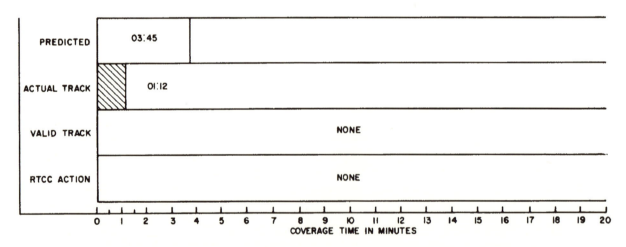

Figure 3-73. HTV USB Tracking Coverage, Reentry -- CSM

3.16.3.2 VHF

3.16.3.3 Telemetry Computer. HTV has no capability.

3.16.4 COMMAND

No capability for command exists.

3.16.5 AIR-TO-GROUND COMMUNCIATIONS

3-92

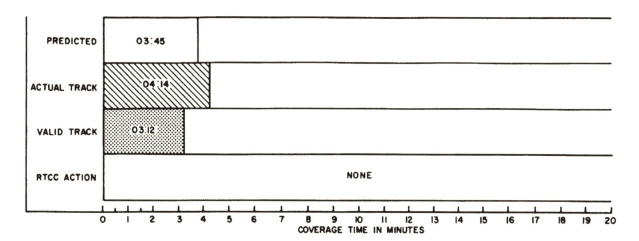

Figure 3-74. HTV CAPRI Tracking Coverage, Reentry -- CM

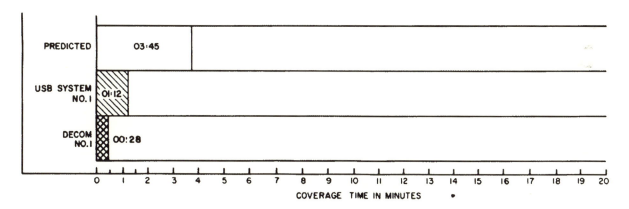

Figure 3-75. HTV Telemetry Coverage, Reentry -- CM

3.17 MADRID/MADRID WING (MAD/MADX) (Figures 3-76 through 3-81)

3.17.1 GENERAL (Figures 3-76 through 3-78)

MAD and MADX provided support during all mission phases with the exception of EO. MAD provided support, active and passive, of the CSM during TLC, the lunar phase, TEC, and approximately 10 minutes support of the LM on LO 27. MADX also provided both active and passive support of the CSM during TLC, the lunar phase, and TEC. In addition, the wing station provided active and passive support of the lunar surface activity and approximately 4 minutes of passive IU support during TLC.

Acquisition messages were a problem at both stations. Messages were received late, sometimes were identical, had overlaps or gaps in the time, or had angle offset errors of up to 0.45 degree. During TLC at 06:02:00 GET, all TTY receive circuits were out for approximately 90 minutes. MAD's messages were exhausted while actively supporting which necessitated a change in mode to enable automatic tracking. In order to alleviate this problem at MADX, a procedure was devised in real time to maintain valid program track. A 30-minute time offset in the APP was added to the last valid acquisition message that was received and angle offsets were applied so that the command angles corresponded to the real declination angle. GWM has noted earlier that the primary lateral motion of a spacecraft in TLC, lunar, and TEC phases

3-93

of lunar missions is the relative motion caused by the rotation of the earth. Any processed 29-point acquisition message for this period reflects primarily this earth-rotational motion, and thus a given tape can be used for various track periods simply by empirically adding the appropriate angle or time bias. Since declination angles remain fairly constant on a lunar mission, this solution may prove to be beneficial to other stations on future missions.

During LO 2 (77:28:00 GET), the correlation voltage from the R/E to the systems monitor was not recorded due to an isolation amplifier failure. Replacement of the amplifier restored the recording capability.

Figure 3-76. MAD/MADX Tracking Station

Figure 3-77. MAD/MADX Mission Support

3-94

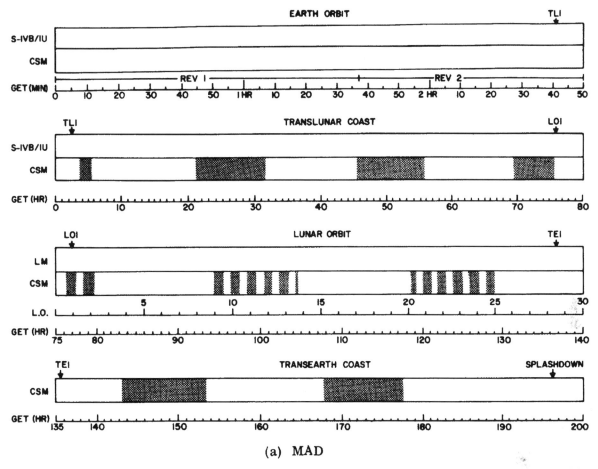

(a) MAD

Figure 3-78. Support Periods

3.17.2 TRACKING

3.17.2.1 <u>Madrid (MAD) USB (Figure 3-79)</u>. Early in TLC (04:02:00 GET), an operator error at GDS occurred during a GDS to MAD HGA handover. The GDS operator used omni-antenna procedures as opposed to those for HGA. However, MAD initiated uplink and was able to acquire two-way lock with no appreciable loss of track.

At 98:08:30 GET, the failure of a timing system distribution amplifier disabled PFS channel No. 1 (3.1.2.4). The problem was resolved prior to AOS for LO 12 by switching to channel No. 2 and resynchronizing the time standards, and had no impact on mission support.

During LO 25 at 124:28:24 GET, severe oscillations were observed in the antenna X-axis. This difficulty was caused by a defective position amplifier in the servo system. Antenna motion in the X-axis was halted, the amplifier replaced, and autotrack maintained for the remainder of the view period.

Shortly after AOS on LO 26, a cooled paramp problem resulted in the loss of autotrack at 126:04:00 GET. The amplifier began oscillating and had to be retuned to restore normal operation (3.1.2.3). The uncooled paramp was placed in service to provide backup voice.

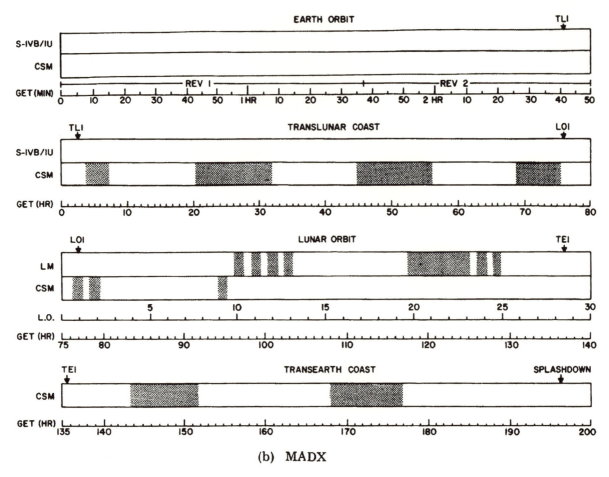

(b) MADX

Figure 3-78. Support Periods (cont)

3.17.2.2 <u>Madrid Wing (MADX) USB (Figure 3-80).</u> Marginal operation of the 1218 computer high-speed tape punch caused a faulty drive tape to be supplied for the APP during TLC at 03:25:30 GET (3.1.2.1).

At 45:46:37 GET, the station's Doppler data was biased by an incorrectly selected frequency synthesizer. After approximately an hour, the operator error was discovered and corrected by selecting the proper frequency.

The high ambient temperature in the hydro-mechanical building was the cause of problems at MADX. At 47:10:00 GET, the wing station transmitters were disabled when the main circuit breaker of the 400-Hz generator tripped. Approximately an hour later at 48:18:10 GET, an ac overcurrent relay tripped on the beam high-voltage generator, again disabling the transmitters. This resulted when MADX raised the beam voltage to maximum limits. Uplink capability was interrupted on both occasions and an unscheduled handover to the prime station was performed.

At 69:15:00 GET, large range jumps of several thousand meters were observed in the ranging data. Troubleshooting revealed a defective PC card (FD-10, slot 13) in the readout register. This card provided bits 12 and 13 of the 30-bit range. Replacement of the card resolved the problem and no further difficulty was encountered.

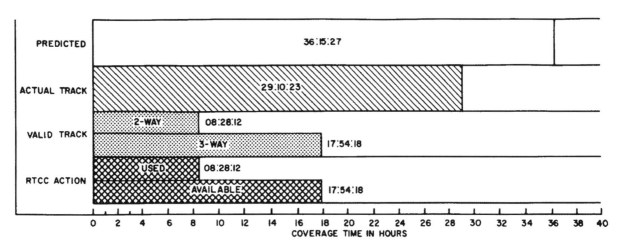

(1) TLC -- CSM

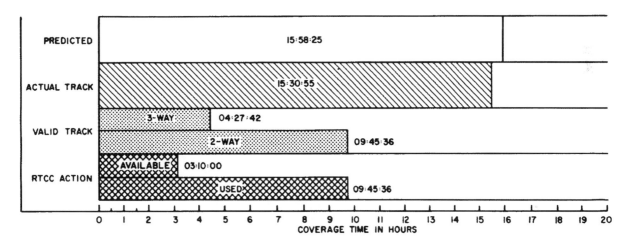

(2) Lunar Phase -- CSM

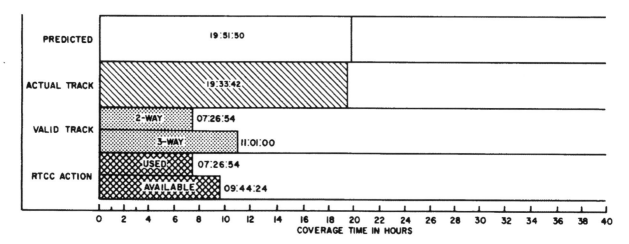

(3) TEC -- CSM

(a) System No. 1

Figure 3-79. MAD USB Tracking Coverage

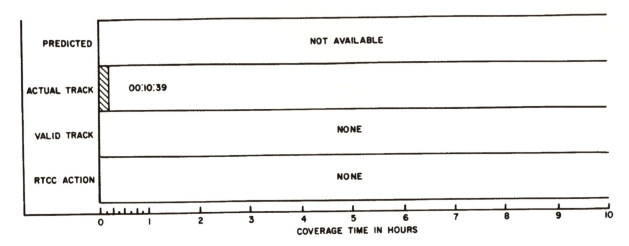

(b) System No. 2, Lunar Phase -- LM

Figure 3-79. MAD USB Tracking Coverage (cont)

About 20 minutes of TDP data was erroneously flagged "bad" during TLC, beginning at 73:43:00 GET, due to an operator error. The error was corrected by resetting the GOOD/BAD data switch to the GOOD position.

During TEC at 144:09:20 GET, 20 seconds of uplink capability was lost when an arc in the high-voltage power supply inhibited the transmitter beam voltage. Uplink capability was restored when the beam voltage was reset (3.1.2.2).

3.17.2.3 Radar. MAD and MADX do not have C-band capability

3.17.3 TELEMETRY (Figure 3-80)

3.17.3.1 USB. All S-band telemetry data processing was accomplished at the prime station; therefore, wing data was transmitted to the prime station via microwave for processing. Minor data losses were encountered when the CSM omni antenna was used for transmission, when the CSM initiated PTC operations during the coast phases, and when bit rate changes were performed (3.1.3.1).

Valid EKG data from MAD was not received at MCC during TLC at 46:58:00 GET. The problem was determined to be a patching error on station which had the EKG parameters reversed. MAD corrected the configuration as soon as MCC informed the station of the problem.

Decom No. 1 encountered difficulty when processing LM data for LO 13 at 100:16:20 GET due to a momentary failure of the wideband signal conditioner (3.1.3.3). Thirty seconds after the problem occurred, the input to the decom was switched to the narrowband signal conditioner. No reason for the malfunction was noted and the unit was able to support the next real-time requirement. There is the possibility that the problem may have been a recurrence of one experienced prior to liftoff when the format selection function became disabled. At that time, cycling of the function switches restored normal operation.

Approximately 14 seconds of LM data was lost during lunar liftoff due to degraded received signal (3.1.3.2).

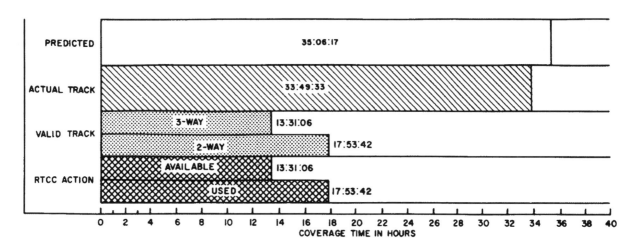

(1) TLC -- CSM

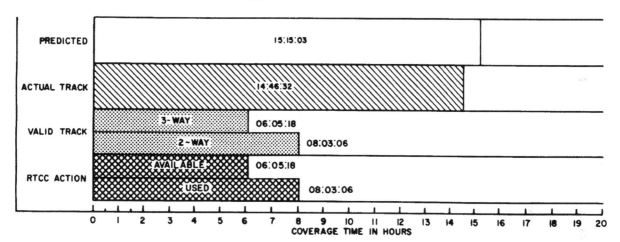

(2) Lunar Phase -- LM

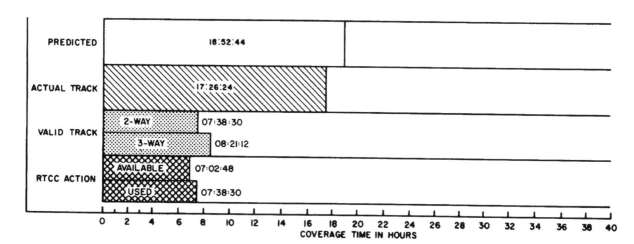

(3) TEC -- CSM

(a) System No. 1

Figure 3-80. MADX USB Tracking Coverage

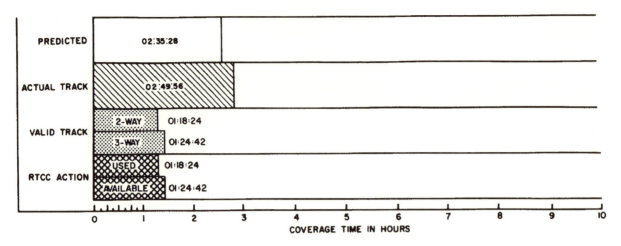

(b) System 2, Lunar Phase -- CSM

Figure 3-80. MADX USB Tracking Coverage (cont)

Several seconds of LM data was lost during LO 26 at 126:42:00 GET due to an SDDS operator error; the incorrect data demod bandwidth was selected. This data was critical and MCC selected ACN's data until the problem was corrected.

A 48-minute loss of CSM downlink occurred during TEC due to a procedural error at MCC. (3.1.1.4).

3.17.3.2 VHF. MAD and MADX have no VHF capability.

3.17.3.3 Telemetry Computer. The telemetry data processor faulted seven times during the mission period. The first four faults occurred prior to launch. Investigation revealed that the faults were caused by a defective inhibit current diverter at location A13J49C. The three remaining faults occurred at 45:16:00 GET during TLC, and at 170:00:00 and 170:03:00 GET during TEC. These faults resulted in losses of high-speed 2.4-kbps data totaling 6 minutes and 38 seconds. The cause of the faults is undetermined; however, faulty memory is suspected. Telemetry computer faults were common problem experienced by other MSFN stations during the mission (3.1.3.4).

A computer problem occurred between 24:13:00 and 25:18:00 GET when MAD was reading fill data for the CM computer words 45B and 95B. Investigation revealed an extra bit at address 10256 which was corrected by reloading HSD format 007 from the system tape.

During LO 11 (95:40:00 GET), the telemetry computer went into a loop and was recovered by the automatic fault recovery program with no data losses. Analysis indicated that a problem existed between channel 14 GMT TP input and channel 15 PAM input. The problem was such that an ID active was established on channel 15 with no PAM input. This problem was bypassed by setting the 1299 switch for GMT TP in the neutral position when not in use. Investigation revealed the cause of the problem to be a missing wire in chassis A1 between 44G8 and 39E7.

3.17.4 COMMAND

Several failures of the command MTU at MAD occurred resulting in RTC history losses. During the early portion of TLC at 05:13:00 GET, an MTU failure occurred and bad

3-100

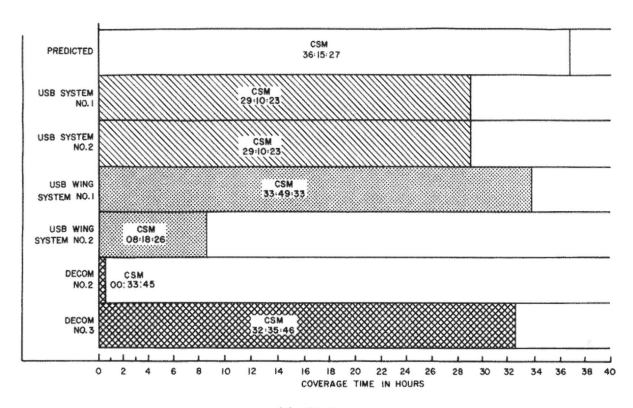

(a) TLC

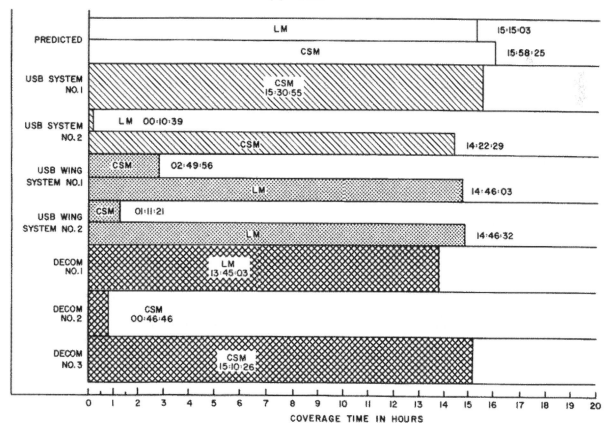

(b) Lunar Phase

Figure 3-81. MAD/MADX Telemetry Coverage

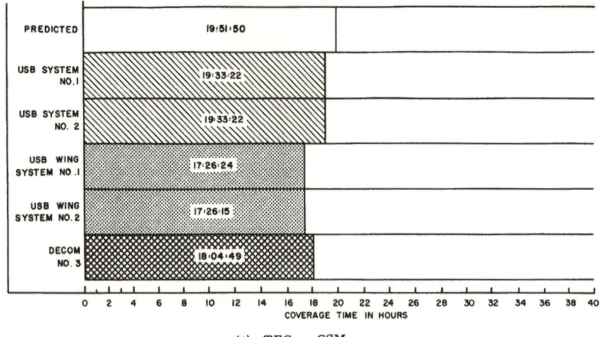

(c) TEC -- CSM

Figure 3-81. MAD/MADX Telemetry Coverage (cont)

history records were written. As a result, real-time and OUCH histories were lost, but command information was later extracted from the HSP and sent to MCC via TTY. At 24:02:04 GET during TLC, MAD again encountered MTU problems with the command histories being lost. The command and telemetry MTU's were interchanged, but the problem still existed.

A load uplink 1001 revealed three bad tape recorders. After the MTU reconfiguration, an end-of-file could not be written necessitating a CBARF recovery. Later during TLC at 28:20:00 GET, the Instrumentation Communications Officer at MCC initiated a RTC execute while MAD was taking a high-speed command history. The history was aborted due to the uplink and was later attempted unsuccessfully. The MTU failures were eventually traced to bit 14 being set during the read operations. This caused an output timing error to occur in the status register and blank data to be written over the histories. All output flip-flops and output amplifier cards for bit 14 were replaced along with the output timing and output acknowledge flip-flops. No further MTU problems were encountered after completing these changes. Failure of the command MTU was also encountered by MER, (4.1.4.3).

3.17.5 AIR-TO-GROUND COMMUNICATIONS

On several occasions the ComTech at MCC queried MAD to determine the cause of distortion on GOSS conference. The problem existed at several stations; however, GDS and MAD reported that the distortion was believed to have been caused by the VOGAA (3.1.5). The VOGAA is suspected of having caused distortion at several other stations and the problem is being investigated by MFED.

At 126:04:00 GET during LO 26, a cooled paramp problem caused a loss of downlink signals (3.17.2.1). Analysis of the voice tapes revealed that the CSM downlink signal was of insufficient level to hear, and the noise level was very high after the uncooled paramp was placed in service. This was the condition as monitored at the output of the SDDS before the signal was supplied to the input of the VOGAA. There was no output from the VOGAA because the noise level exceeded the voice signal level, thus suppressing both signal and noise. At the time, ACN indicated good downlink was being received with the VOGAA bypassed. MCC informed MAD that both ACN and MIL were receiving good backup voice from the CSM; however, 14 minutes elapsed before MCC switched to ACN.

3.18 MERCURY (MER) (Figures 3-82 through 3-87)

3.18.1 GENERAL (Figures 3-82 through 3-84)

MER's primary requirement was to provide CSM S-band support during the critical S-IVB TLI burn until engine cutoff (02:44:16 until 02:50:03 GET). The support provided did not satisfy this requirement. Handover of the S-band carrier from RED to MER was delayed until the last moment, and contingency handover procedures were put into effect by MSC/RTC at Houston because of MER's malfunctioning command computer.

No telemetry computer problems were encountered. A/G communications (VHF) were very good.

Figure 3-82. USNS Mercury

3-103

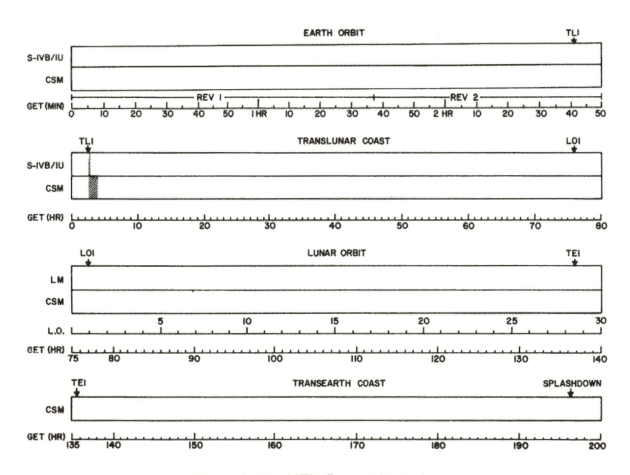

Figure 3-83. MER Mission Support

Figure 3-84. MER Support Periods

3.18.2 TRACKING

3.18.2.1 USB (Figure 3-85). MER was required to provide S-band tracking support during TLI. This primary requirement was generally unsuccessful. Handover of the S-band carrier from RED to MER was delayed by MSC because of a command computer failure.

Although the command computer problem was not corrected, MSC initiated contingency handover from RED to MER because RED was about to experience LOS. A few seconds after handover was accomplished, the command computer was operational and MER was

3-104

"go" for CSM command at 02:47:43 GET. At this time, signals were noisy and numerous tracking losses occurred due to attenuation of the S-band carrier caused by the S-IVB flame plume. As a result, USB metric data was not flagged two-way and the data was rejected by MSC. HAW had AOS at 02:49:04 GET and MSC, dubious of MER support capabilities, initiated contingency handover procedures from MER to HAW. The early handover was successfully accomplished at 02:49:46 GET. All MER data received from this point on was ignored by MSC.

MER had a flexure monitor equipment failure prior to the pass due to a shorted transistor. Upon replacement, the equipment was declared operational. However, the same transistor shorted again during the pass at 02:45:00 GET. The flexure monitor contingency procedures contained in the NOD were exercised. But an attempt to load the misalignment coefficients in accordance with ISI 043, while the computer was cycling in real time, caused a CDP fault. MER's MMR stated that the problem was due to software within the CDP, but the exact nature was not given (3.1.2.5).

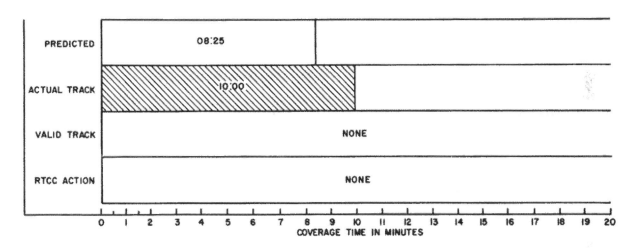

(a) System No. 1, EO -- CSM

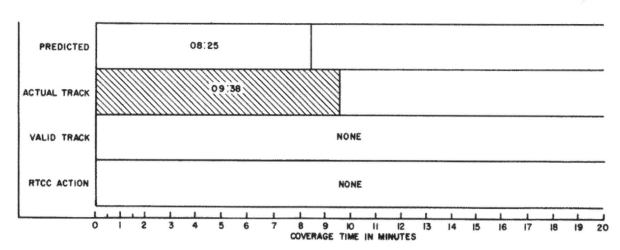

(b) System No. 2, EO -- IU

Figure 3-85. MER USB Tracking Coverage

3.18.2.2 <u>Radar (Figure 3-86)</u>. During TLI, the S-IVB/IU was acquired 2 minutes and 36 seconds later than predicted. As a result, MER provided only limited TLI support. Acquisition was not accomplished until the TLI burn was nearly completed. This was caused by MER's failure to comply with ISI 053 which directed the use of beam intercept

3-105

techniques for acquisition. Instead, MER initially elected to use the IRV to acquire. When this failed, the S-IVB/IU was finally acquired by slaving to the USB and using manual offset. An additional loss of data resulted from phasing problems with another radar. All MER C-band data was inhibited by MSC RTCS because of the higher priority for USB metric data.

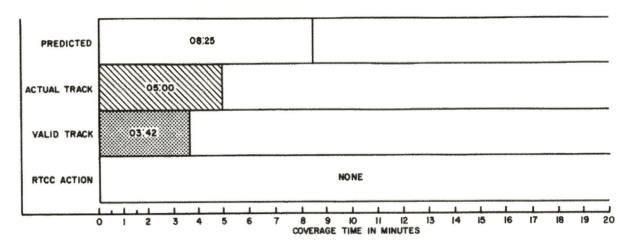

Figure 3-86. MER FPS-16 Tracking Coverage, EO -- IU

3.18.3 TELEMETRY (Figure 3-87)

3.18.3.1 USB. The received signals from the CSM were erratic and noisy during the TLI burn due to S-band tracking problems and flame attenuation previously mentioned. Therefore, the primary requirement of providing CSM telemetry processing during the burn and at engine cutoff did not satisfy mission requirements.

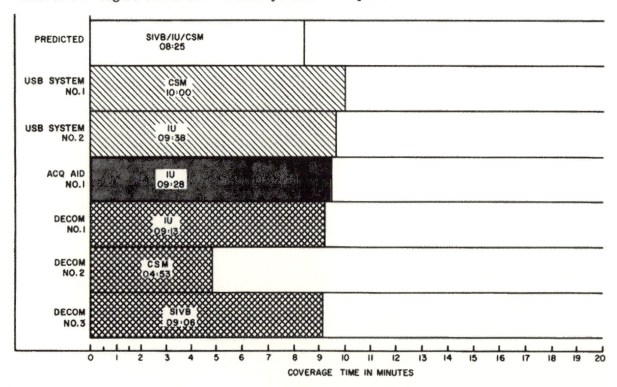

Figure 3-87. MER Telemetry Coverage, EO

3-106

3.18.3.2 <u>VHF.</u> Acq aid 4-1 provided IU and S-IVB VHF telemetry support during TLI and the first few minutes of TLC with minimal data losses.

3.18.3.3 <u>Telemetry Computer</u>

3.18.4 COMMAND

During prepass testing at 02:15:00 GET, the MTU failed when the command computer attempted a write operation, causing loss of command history and recovery capability. The problem was corrected by replacing a defective PC card in the MTU and the equipment was restored to an operational status at 02:38:00 GET. Command MTU failures were also noted at GDS, (3.1.4.3). The command computer faulted at 02:39:00 GET. Technicians manually reloaded the computer after two attempts using CBARF failed. Command capability was restored at 02:47:08 GET, 17 seconds after contingency handover procedures were initiated by MSC due to approaching LOS at RED. MSC delayed handover until the last possible moment and minimized mission impact. (Refer to paragraph 3.1.4.1 for other stations having command computer faults.) The flexure monitor problem caused MSC to initiate early handover to HAW at 02:49:46 GET. Consequently, MER did not provide command coverage at engine cutoff, which was a primary mission requirement.

3.18.5 AIR-TO-GROUND COMMUNICATIONS

3.19 <u>MERRITT ISLAND (MIL) (Figures 3-88 through 3-93)</u>

3.19.1 GENERAL (Figures 3-88 through 3-90)

MIL actively and passively supported the CSM, and passively supported the IU during launch and EO revolution 2. All remaining support was passive and included the IU during early TLC, the CSM during TLC and TEC, the CSM and LM during the lunar phase, and the LM during a portion of its lunar stay.

At launch, tracking difficulty was experienced because of excessively strong signals. A number of tracking modes were rapidly tried in order to obtain the best possible metric data.

Several stripchart recorder problems occurred during TLC support periods. At 08:17:00 GET, the autotrack event pen for systems monitor Brush recorder No. 3 became clogged with ink and was inoperative. Autotrack event data was lost for 7 minutes until the pen was replaced. Later at 52:17:00 GET, the channel 8 pen (range correlation voltage) of this same recorder failed. Fortunately, this failure caused no problem because the tracking was three-way at this time, and when in this tracking mode, the range correlation data is invalid. During a passive tracking period, systems monitor recorder No. 4 lost 24 minutes of recorded AGC and angular data at 60:55:00 GET when the recorder ink supply motor failed and the ink supply became depleted. A problem with the 100-channel event recorder occurred at 35:02:00 GET when a pulley drive wheel retaining screw failed and the recorder would not operate at 2 mm/second. The recorder speed was changed to 200 mm/second and recording was resumed. One and a half minutes of recorded events data was lost because of this failure. At 102:53:00 GET while tracking the CSM on LO 14, systems recorder No. 4 channel 1 pen failed because of a broken tension band. Approximately 29 minutes of X-angle recorded data was lost during the installation of a new band. The 100-channel event recorder stopped during during LO 30 at 133:59:00 GET because a manufacturer's splice in the paper roll jammed the recorder. Recording was resumed after 1 minute and 10 seconds of recorded events data was lost. Several MSFN stations encountered systems monitor problems (3.1.1.5).

3-107

An operator error at GDS prevented correct time correlation tests at 56:27:00 GET (3.1.1.2).

While passively tracking the CSM during LO 3, a loss of downlink signal was encountered due to an unscheduled handover from GDS to HAW (3.1.1.3).

A 48-minute loss of CSM downlink occurred during TEC due to a procedural error at MCC (3.1.1.4).

No command computer problems were experienced during support activity. USB A/G communications supported passively without experiencing any problems.

3.19.2 TRACKING

3.19.2.1 USB (Figure 3-91). At 56:19:00 GET, momentary oscillations on channel 1 of the PFS caused an intermittent 1-MHz timing signal to the R/E. An immediate selection of distribution amplifier channel 2, followed by reselection of channel 1, cleared the problem without any loss of data (3.1.2.4).

Following postpass calibrations during the lunar phase at 88:08:00 GET, the antenna was observed entering prelimits 4 degrees early and exiting late. The problem was caused by dirty and pitted contacts on the Y-axis south limits. Cleaning and burnishing the contacts restored the system to operational status.

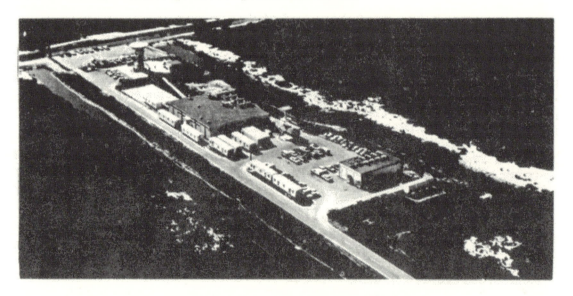

Figure 3-88. MIL Tracking Station

SYSTEM	EARTH ORBIT		TRANSLUNAR COAST				LUNAR PHASE																															TRANS-EARTH COAST		
	REV		DAY				REVOLUTION																															DAY		
	1	2	1	2	3	4	1	2	3	4	5	6	7	8	9	10	11	12	13	14	15	16	17	18	19	20	21	22	23	24	25	26	27	28	29	30	31	7	8	9
USB NO. 1	1	1	1	1	1	1	1	1	1	1	1							4	4	1	1	1	1								1	1	4	4				1	1	
USB NO. 2	5	5	1/5	1	1	1	1	1	1	1	1							1	1	4	4	4	4								4	4	1	1	1	1		1	1	
ACQ AID NO. 1	5																																							
ACQ AID NO. 2	5	5	5																																					
DECOM NO. 1	(5)	(5)	(5)/5															4	4	4	4	4									4	4	4	4						
DECOM NO. 2	(6)	(6)	1	1	1	1	1	1	1	1	1							1	1	1	1	1	1								1	1	1	1	1	1				
DECOM NO. 3	1	1	1	1	1	1	1	1	1	1	1							1	1	1	1	1	1									1	1	1	1	1		1	1	
DECOM NO. 4	(9)								1									4	4													4	4							
DECOM NO. 5																																1								

1=CSM 4=LM 5=IU 6=S-IVB 9=S-II

O=VHF TELEMETRY PROCESSED

▒=LM LUNAR STAY

Figure 3-89. MIL Mission Support

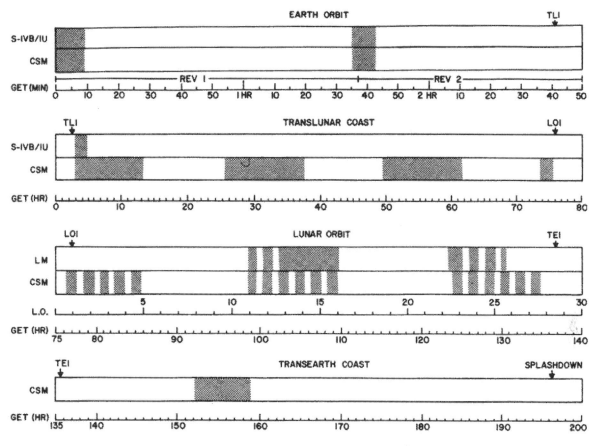

Figure 3-90. MIL Support Periods

Three times during support, the TDP tape punch jammed, but the problem was solved each time by clearing the punch. In each case no data was lost because communications personnel were duplicating it.

3.19.2.2 Radar. MIL has no capability.

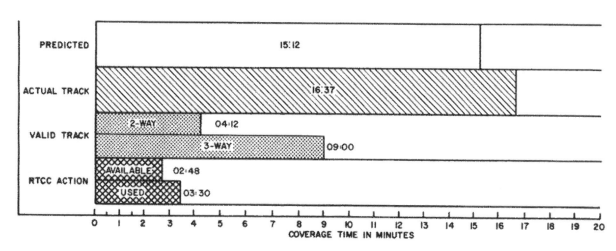

(1) EO -- CSM

(a) System No. 1

Figure 3-91. MIL USB Tracking Coverage

3-109

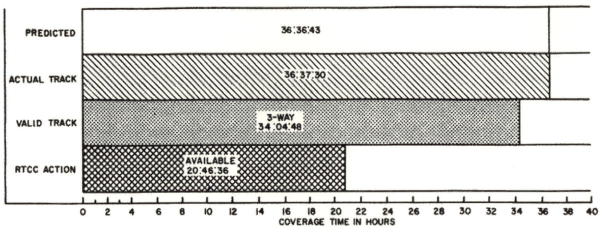

(2) TLC -- CSM

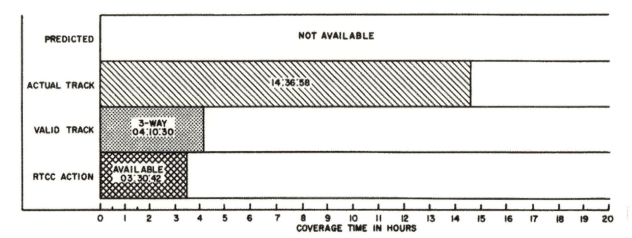

(3) Lunar Phase -- LM

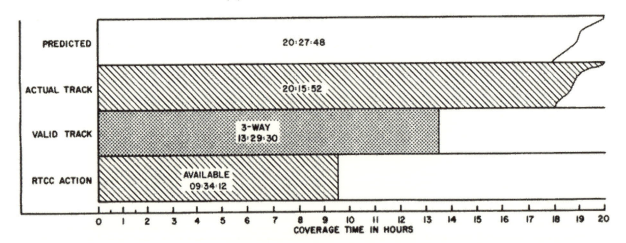

(4) Lunar Phase -- CSM

(a) System No. 1 (cont)

Figure 3-91. MIL USB Tracking Coverage (cont)

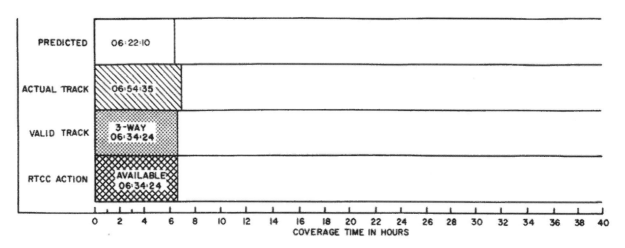

(5) TEC -- CSM

(a) System No. 1 (cont)

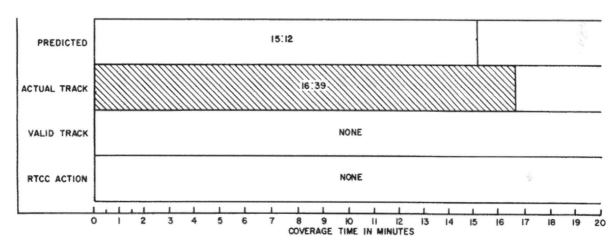

(1) EO -- IU

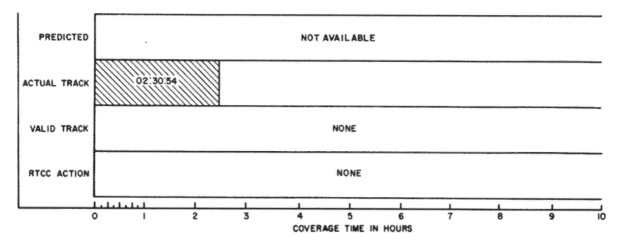

(2) TLC -- IU

(b) System No. 2

Figure 3-91. MIL USB Tracking Coverage (cont)

3.19.3 TELEMETRY (Figures 3-92 and 3-93)

3.19.3.1 USB. Real-time data, both HBR and LBR, from the CSM was processed during EO, TLC, the lunar phase, and TEC. The IU real-time data was processed early in TLC. HBR data from the LM was processed during the lunar phase at the time of descent to lunar landing, including a portion of the lunar stay. LM data was also processed from lunar liftoff to docking with the CSM. One CSM dump was attempted during the lunar phase, but the dump was unsuccessful because of insufficient signal levels.

Data losses experienced during support periods were attributed to signal levels below decom threshold, use of the omni antenna on the spacecraft, the PTC mode of flight during TLC and TEC, unfavorable aspect angle of the spacecraft antenna, and bit rate changes (3.1.3.1).

During the first 7 minutes of launch, CSM PM data was routed through the FM data demod. This configuration was selected because the FM demod does not require a phase lock loop in order to process data, whereas the PM demod is dependent upon the correct phasing to maintain lock. During the initial part of the launch phase, PM downlink data is phase-shifted due to the booster flame and usually causes the PM demod and, subsequently, the decom, to drop lock. In this new configuration, continuous CSM PM data processing was assured even though a 6- to 7-dB loss in signal strength could be expected. Since the downlink signal level is extremely strong during launch because of the close proximity of the vehicle, this loss was considered negligible. MIL tested the new configuration during AS-505, and the net result indicated that better support could be provided using the FM demod arrangement during the early portion of launch. Also, it afforded the PCM operator an option to select either PM or FM channel inputs to the decom. Since the need for good PM USB data is mandatory during launch, this configuration is expected to be used during future missions tor launch support. Figure 3-92 is a block diagram depicting data flow when PM data is routed through the FM/ demod and the switching capability.

Approximately 14 seconds of LM data was lost during lunar liftoff due to degraded received signal (3.1.3.2).

3.19.3.2 VHF. Telemetry from the S-IVB and IU was processed during launch and EO revolution 2, and the IU was supported during early TLC. The S-II vehicle telemetry was supported during launch. There were no equipment or operator problems experienced during support activity.

3.19.3.3 Telemetry Computer. MIL used system B for processing data with system A available for backup support.

On launch day at 04:48:00 GMT, system B experienced problems processing the sector word dump special function set downlinked from the launch vehicle digital computer. The associated software problem was encountered when an external interrupt on EMU channels containing an invalid ID was randomly being received before the last monitor interrupt occurred. The external interrupt containing an invalid ID would incorrectly cause some areas of core to be cleared, which resulted in the improper processing of the special function set. An errata (T8) was available which would have corrected this problem. However, sufficient time was not available for MCC to validate and forward the errata to the station prior to support. This errata will be incorporated in the telemetry program for AS-507. This problem did not occur during launch, therefore, no mission impact resulted.

During the mission CADFISS, MIL experienced problems in remotely selecting telemetry formats (3.1.3.4).

3-112

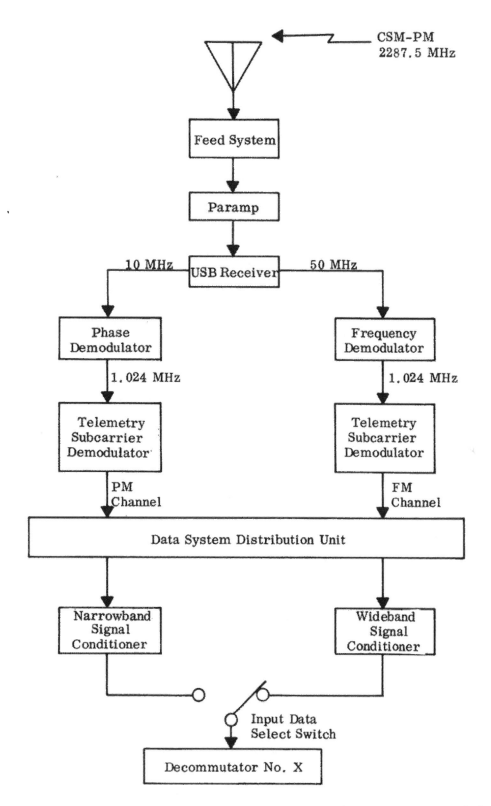

Figure 3-92. MIL USB CSM PM Downlink Telemetry Configuration -- Launch Only.

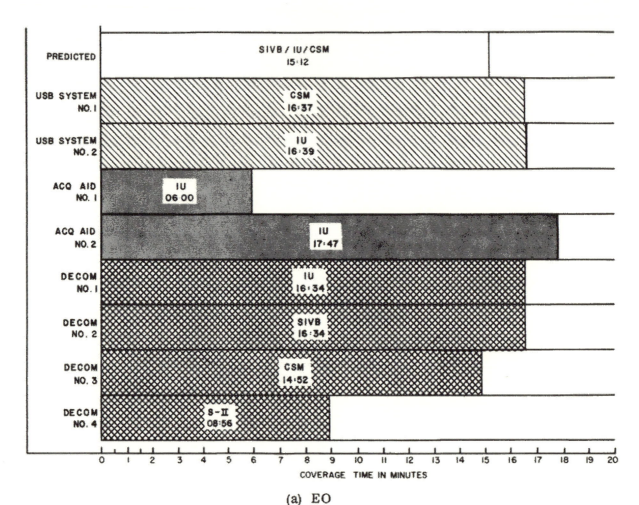

(a) EO

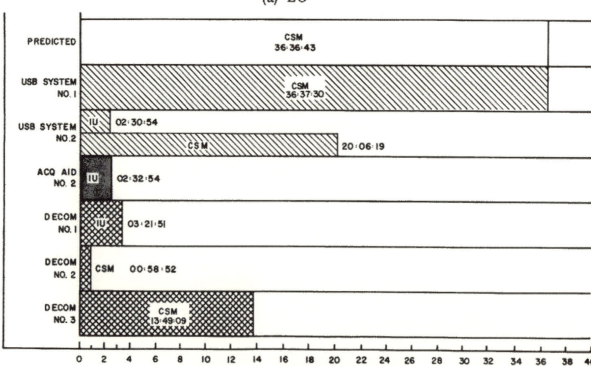

(b) TLC

Figure 3-93. MIL Telemetry Coverage

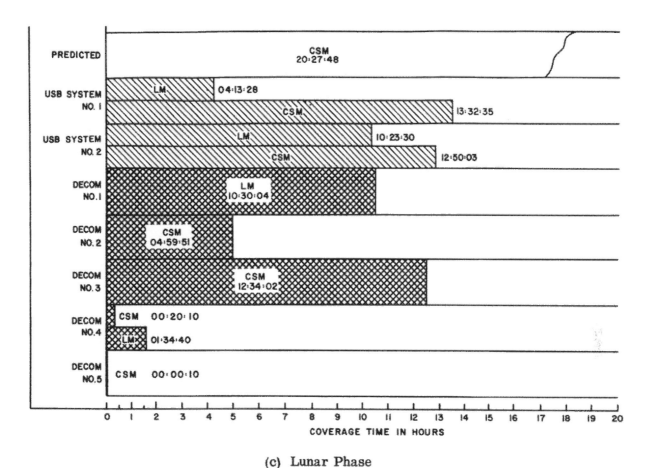

(c) Lunar Phase

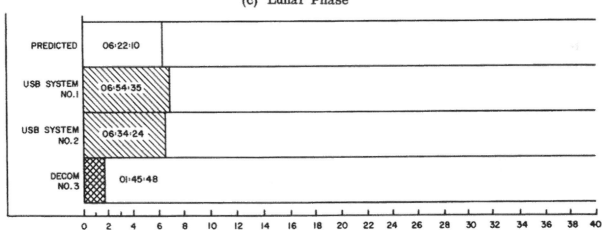

(d) TEC -- CSM

Figure 3-93. MIL Telemetry Coverage (cont)

3.19.4 COMMAND

3.19.5 AIR-TO-GROUND COMMUNICATIONS

3.19.5.1 USB

3.19.5.2 VHF. Acq aid No. 1, configured for VHF A/G voice, failed to track the CSM during the initial portion of the launch phase. The failure was attributed to a defective comparator in the antenna feed system. Water had entered during a

3-115

heavy rain storm and shorted the comparator. The problem occurred sometime between the final alignment at T-3 minutes and 10 seconds and liftoff, and became apparent only after two attempts to autotrack failed. A degraded autotrack was successfully accomplished at approximately T+1 minute and 23 seconds. No data was lost and A/G communications were supported without mission impact. After supporting launch, acq aid No. 1 was declared "red." The defective comparator was replaced with a new unit and the system was restored to an operational status at 49:16:00 GET. However, VHF support was terminated prior to this time. All MIL VHF requirements until support termination were accomplished using acq aid No. 2.

3.20 REDSTONE (RED) (Figures 3-94 through 3-99)

3.20.1 GENERAL (Figures 3-94 through 3-96)

RED was assigned a TSP of 2.25 degrees south and 166.8 degrees east for its primary mission requirement of providing CSM and IU support during the end of EO revolution 2, including the beginning of TLI. A secondary requirement of providing CM reentry support necessitated the ship's getting underway, after successful completion of TLI support, and repositioning at TSP 8.0 degrees south and 164.0 degrees east.

Following TLI support, a successful data transfer was accomplished from RED to ARIA 8, which not only expedited delivery of TLI data to MSFC, but also permitted an earlier departure by RED for the reentry TSP.

The telemetry computer, command, and A/G communications areas did not encounter any problems during the mission.

Figure 3-94. USNS Redstone

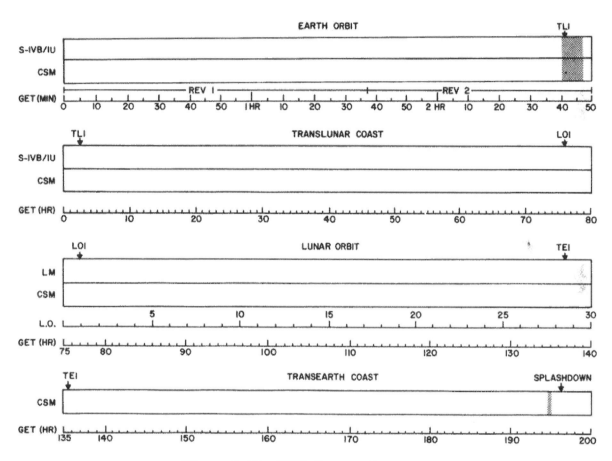

Figure 3-95. RED Mission Support

Figure 3-96. RED Support Periods

3.20.2 TRACKING

3.20.2.1 USB (Figure 3-97). The primary activity during the premission stopover at Kwajalein was to ensure optimum phasing of the S-band system. TETR-B tracking had indicated the phasing to be invalid. The system was rephased according to readings previously established at Pearl Harbor and subsequently validated again by tracking TETR-B. Unfortunately, the system cannot be readily phased at sea; therefore, it inhibits crew capability to ensure the integrity of front-end alignment and repair action.

3-117

Just prior to launch, the flexure monitor laser assembly failed. The defective components causing the failure were replaced at sea, but operation was not satisfactory due to the lack of dockside availability to perform roller-path alignment. RED implemented the NOD contingency procedure by zeroing out misalignment coefficients and reduced mission impact to a minimum (3.1.2.5).

During precalibrations for EO revolution 2, fluctuation in the gain of the main paramp was observed as a result of ac power surges. The power source stabilized before tracking operations began and no data was lost.

A contingency late handover during TLI resulted in RED providing more active coverage than planned, in that active two-way lock on the CSM was maintained 1 minute longer. A command computer MTU failure aboard the MER brought about this extended requirement. RED provided good quality metric tracking data and monitored the S-IVB ignition and early portion of the TLI burn, as required.

During the final minutes of TEC at 194:44:19 GET, RED provided coverage of the critical CM/SM separation maneuver. Metric tracking data was excellent. Prior to this view period while conducting prepass checks, several CDP buffers failed when S-band timing signals were lost due to a patching problem. Necessary timing signals were repatched to the CDP to restore normal operation before the start of tracking.

3.20.2.2 Radar (Figure 3-98). During EO revolution 2, the FPS-16 provided nominal tracking support of the IU, a primary mission requirement. A secondary mission requirement of the FPS-16 was to provide metric tracking of the SM after CM/SM separation during reentry. Acquisition information provided RED indicated that the CM would precede the SM and that the radar should track the trailing target. Such was not the case, because the SM led the CM and, consequently, RED tracked the wrong vehicle.

3.20.3 TELEMETRY (Figure 3-99)

3.20.3.1 USB. The Redstone provided S-band telemetry support for the CSM and IU during EO revolution 2, including TLI and the CSM during reentry. Some data losses occurred while performing a contingency handover to MER during the TLI burn.

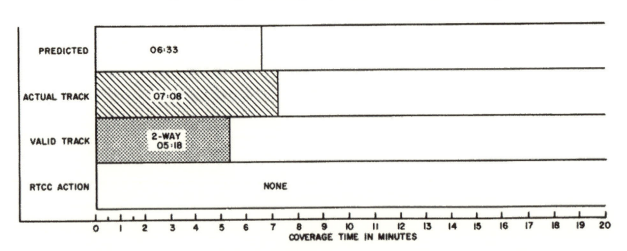

(1) EO -- CSM

(a) System No. 1

Figure 3-97. RED USB Tracking Coverage

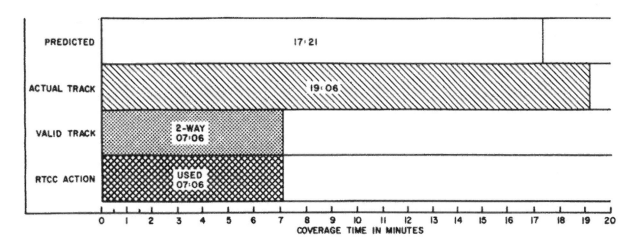

(2) Reentry -- CM
(a) System No. 1 (cont)

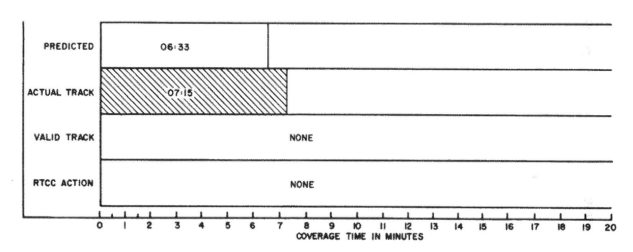

(b) System No. 2, EO -- IU

Figure 3-97. RED USB Tracking Coverage (cont)

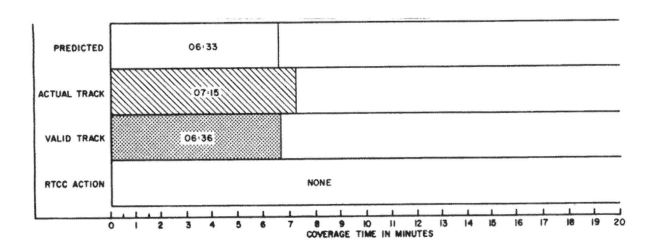

Figure 3-98. RED FPS-16 Tracking Coverage, EO -- IU

3-119

A premission problem with a wideband signal conditioner was resolved prior to launch and caused no difficulty during mission support.

3.20.3.2 VHF. The 4-1 antenna supported the S-IVB and IU during the TLI burn and provided good telemetry data from both vehicles. Two seconds of IU and 55 seconds of S-IVB data was lost because of inadequate signal levels. The CM was tracked for approximately 19 minutes during reentry and good telemetry was provided.

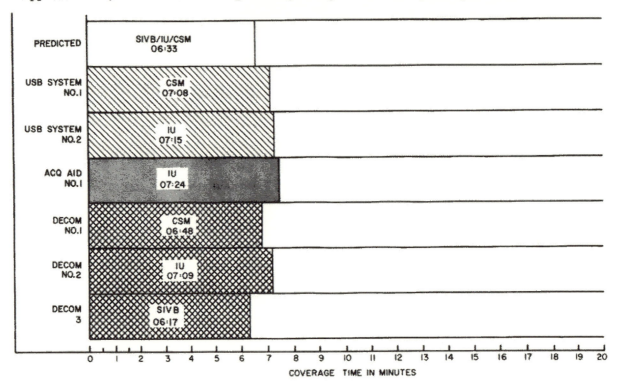

(a) EO

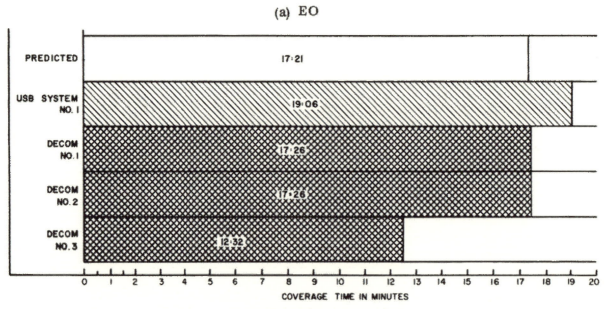

(b) Reentry -- CM

Figure 3-99. RED Telemetry Coverage

3.20.3.3 Telemetry Coverage

3.20.4 COMMAND

3.20.5 AIR-TO-GROUND COMMUNICATIONS

3.21 TANANARIVE (TAN) (Figures 3-100 through 3-104)

3.21.1 GENERAL (Figures 3-100 through 3-102)

TAN supported EO only. No problems were encountered in receiving and recording VHF telemetry.

3.21.2 TRACKING

3.21.2.1 USB. TAN has no S-band capability.

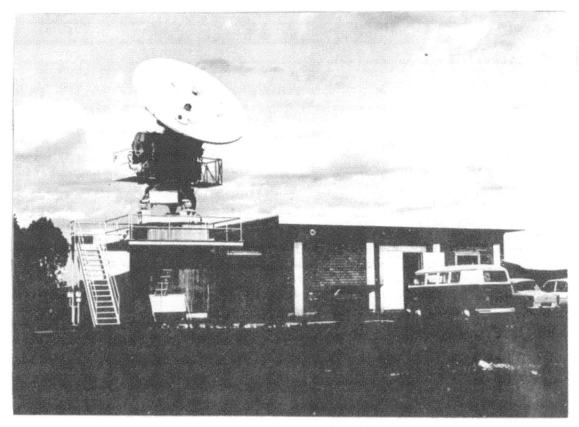

Figure 3-100. TAN Tracking Station

	EARTH ORBIT		TRANSLUNAR COAST				LUNAR PHASE																															TRANS-EARTH COAST		
	REV		DAY				REVOLUTION																															DAY		
SYSTEM	1	2	1	2	3	4	1	2	3	4	5	6	7	8	9	10	11	12	13	14	15	16	17	18	19	20	21	22	23	24	25	26	27	28	29	30	31	7	8	9
CAPRI	5	5																																						
ACQ AID NO. 1	6	6																																						

5=IU 6=S-IVB

(shaded) = LM LUNAR STAY

Figure 3-101. TAN Mission Support

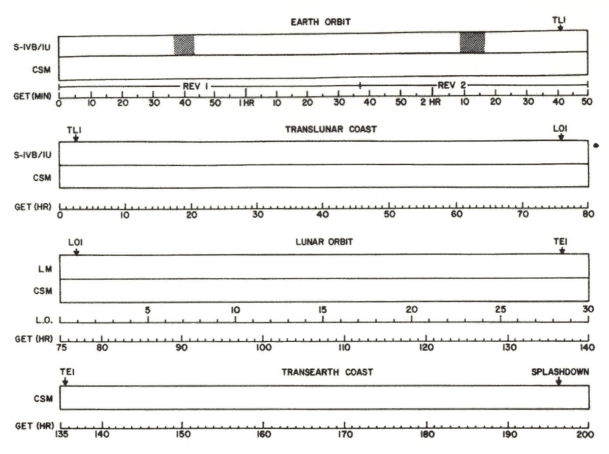

Figure 3-102. TAN Support Periods

3.21.2.2 <u>Radar (Figure 3-103)</u>. The CAPRI tracked the S-IVB/IU during EO. During EO revolution 1, the C-band low-speed tracking data was transmitted in real time, but was not received at GSFC. Because of voice circuit restrictions, TAN was not notified of this fact until after the pass, at which time the data was retransmitted. Postpass transmissions revealed range data to be erroneous for the entire tracking period. The problem was attributed to improper adjustments of find and verify threshold levels, which were apparently corrected before the next support period. The S-IVB/IU was well into EO revolution 2 at an elevation of 15 degrees before the first signals from the beacon were received. The reason for the long delay in acquisition is not known. Improper IDRAN adjustments may have been a contributing factor. The MMR reports that there were no problems during this pass.

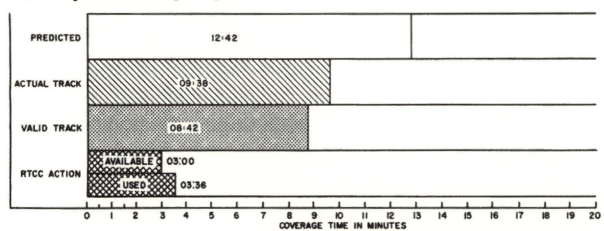

Figure 3-103. TAN CAPRI Radar Tracking Coverage, EO -- IU

3.21.3 TELEMETRY (Figure 3-104)

3.21.3.1 USB. TAN has no USB capability.

3.21.3.2 VHF

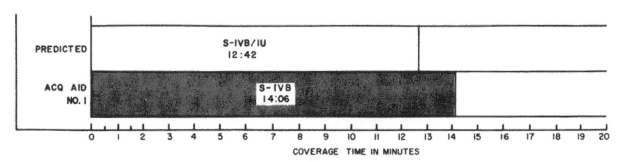

Figure 3-104. TAN Telemetry Coverage, EO

3.21.4 COMMAND

No capability for command exists at TAN.

3.21.5 AIR-TO-GROUND COMMUNICATIONS

During EO revolution 2, a failure on the LDN receive circuit caused a loss of MCC uplink voice. The failure went undetected by the GCC at TAN as he was monitoring all three receive circuits simultaneously and had copied the MCC voice uplink attempts. However, the M&O and ComTech supervisors did not hear the voice uplink attempts because loop No. 1 of the 112A key system, which they were monitoring, was configured to receive on the LDN circuit. The LDN open line was corrected, but not until 1 minute and 15 seconds of voice uplink was lost.

3.22 TEXAS (TEX) (Figures 3-105 through 3-109)

3.22.1 GENERAL (Figures 3-105 through 3-107)

Support requirements were terminated following LO 15 when TEX reconfigured to support EASEP at 106:36:00 GET.

An operator error at GDS prevented correct time correlation tests at 56:27:00 GET (3.1.1.2).

While passively tracking the CSM during LO 3, a loss of downlink signal was encountered due to an unscheduled handover from GDS to HAW (3.1.1.3).

There were no telemetry computer, command, or A/G communications problems encountered.

3.22.2 TRACKING

3.22.2.1 USB (Figure 3-108). During TLC at 25:35:00 GET, the USB PA failed due to a motor generator shutdown. A buildup of carbon deposits on the motor generator field rheostat (used for optimizing power factors) caused arcing and tripped the primary power circuit breaker. This resulted in the loss of uplink capability. The rheostat was cleaned and adjusted, and the PA became operational at 27:06:00 GET (3.1.2.2).

Figure 3-105. TEX Tracking Station

Figure 3-106. TEX Mission Support

Following an unscheduled USB handover from GDS to HAW during LO 3, TEX encountered approximately a 6-minute dropout of CSM downlink (3.1.1.3).

At 100:30:00 GET, the timing outputs were disabled by the failure of a -20 volt trygon power supply in the USB timing system's "A" bank. Both phase "A" and "B" banks dropped to a 0-volt output and could not be returned to an on-line status after the -20 volt power supply in the phase "A" bank was replaced. The interlock circuitry was disconnected in order to restore power to the banks. Upon checking the input ac power, it was observed that the phase "A" and "B" 40-amp circuit breakers were incorrectly tied together mechanically during original installation, and one would not trip without the other. Two single-trip type breakers were installed to enable switchover and the banks were activated. Both "A" and "B" clocks lost synchronization during the failure and had to be resynchronized to provide proper timing signals to other equipment. The system was operational at 103:17:00 GET. During this time there was no requirement for tracking data from TEX and the problem had no adverse effect on the mission.

3.22.2.2 **Radar.** No capability exists at TEX.

3.22.3 TELEMETRY (Figure 3-109)

3-124

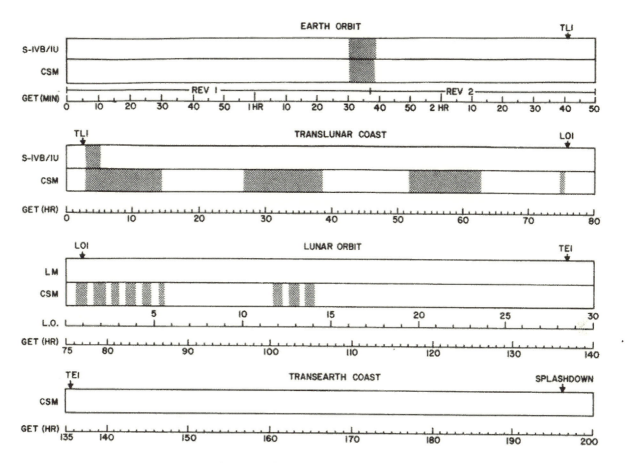

Figure 3-107. TEX Support Periods

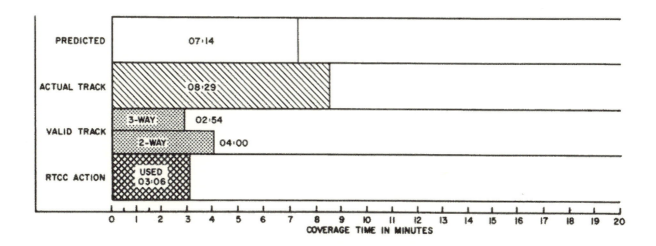

(a) EO -- CSM

Figure 3-108. TEX USB Tracking Coverage

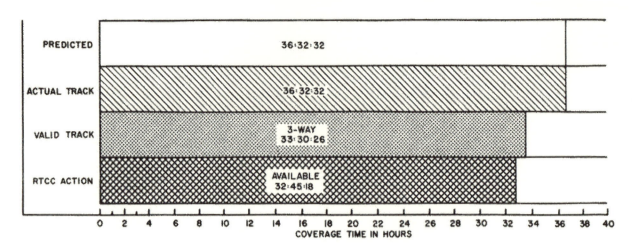

(b) TLC -- CSM

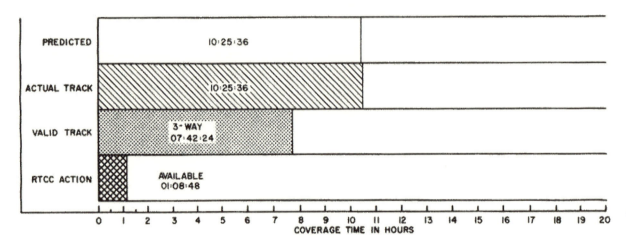

(c) Lunar Phase -- CSM

Figure 3-108. TEX USB Tracking Coverage (cont)

3.22.3.1 <u>USB.</u> Data dropouts were experienced due to unfavorable aspect angle of the spacecraft, the PTC mode of flight, the use of the omni antenna, and low to high bit rate changes (3.1.3.1).

TEX experienced no S-band telemetry problems other than those already mentioned and the approximate 6-minute and 30-second loss due to the handover from GDS to HAW.

3.22.3.2 <u>VHF.</u> A modification (EI 3912) of TEX preamplifiers was incorporated for the support of this mission. New preamplifiers were installed using left- and right-hand circular polarization channel inputs replacing the 0- and 90-degree inputs. The modification resulted in the elimination of the critical matching network and improved system reliability. TEX requirements for VHF telemetry support terminated during the early portion of TLC at 06:26:00 GET when the spacecraft exceeded the system's range capability.

3.22.3.3 <u>Telemetry Computer</u>

3.22.4 COMMAND

3.22.5 AIR-TO-GROUND COMMUNICATIONS

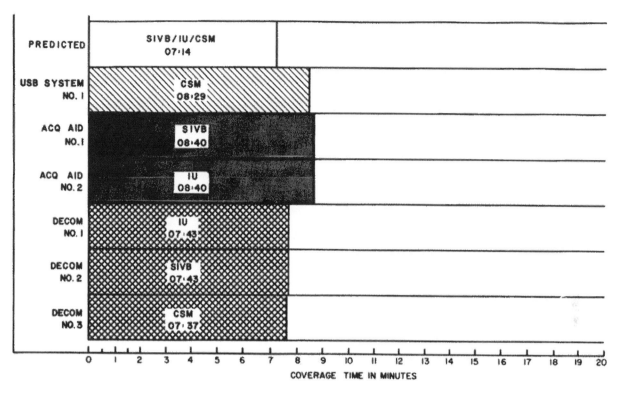

(a) EO

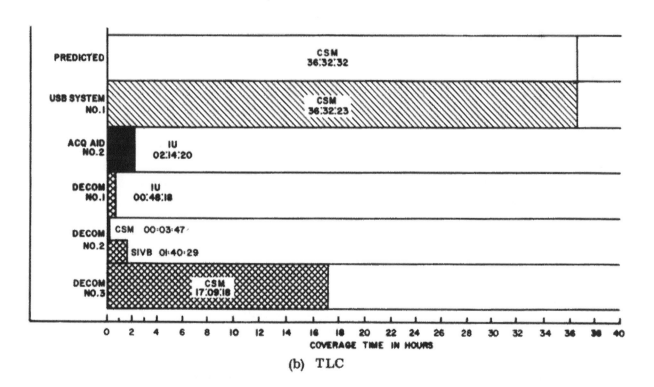

(b) TLC

Figure 3-109. TEX Telemetry Coverage

3-127

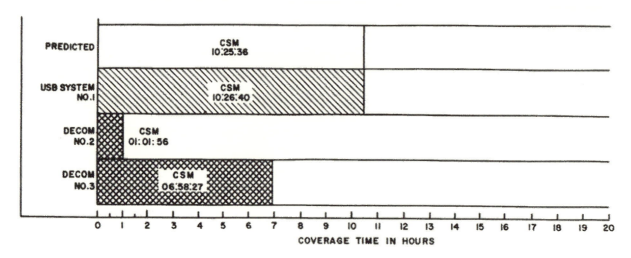

(c) Lunar Phase

Figure 3-109. TEX Telemetry Coverage (cont)

3.23 VANGUARD (VAN) (Figures 3-110 through 3-115)

3.23.1 GENERAL (Figures 3-110 through 3-112)

VAN was assigned a TSP at 28.0 degrees north, 49.0 degrees west to support launch and EO revolution 2. The ship was released after TLI and proceeded to Bermuda for off-loading of data.

There were no VHF telemetry problems encountered during support. The command computer had no hardware or software problems, and A/G communications were supported with no difficulty experienced.

3.23.2 TRACKING

3.23.2.1 USB (Figure 3-113). After the ship departed Port Canaveral, a USB flexure monitor tube failed on July 7. No spare was available and replacement was not possible while at sea. During mission support, VAN used contingency procedures contained in the NOD supplements to zero out misalignment coefficients (3.1.2.5).

VAN did not receive a 29-point acquisition message for launch and locked on a carrier sideband on the initial two-way attempt. This degraded two-way data until successful acquisition of the main carrier.

Postpass analysis of the S-band metric data for both VAN view periods showed the quality of data to deteriorate toward the end of the pass. The problem proved to be minor and of no mission impact but, because no known problem exists on the ship, it is under the investigation of MFED for corrective action if required.

3.23.2.2 Radar (Figure 3-114). VAN FPS-16 encountered a 27-second dropout of data while tracking the IU beacon on EO revolution 2. The actual cause of the data loss is unknown, but it is believed that the radar may have slipped to a side lobe when the vehicle passed within close range of the ship.

3.23.3 TELEMETRY (Figure 3-115).

3.23.3.1 USB.

3-128

3.23.3.2 VHF.

3.23.3.3 Telemetry Computer. At 02:32:00 GET during EO revolution 1, the computer faulted and automatic recovery was successful. Investigation revealed a bad type 2050 PC card. No data was lost because VAN was not supporting at this time (3.1.3.4).

3.23.4 COMMAND

3.23.5 AIR-TO-GROUND COMMUNICATIONS

Figure 3-110. USNS Vanguard

Figure 3-111. VAN Mission Support

3-129

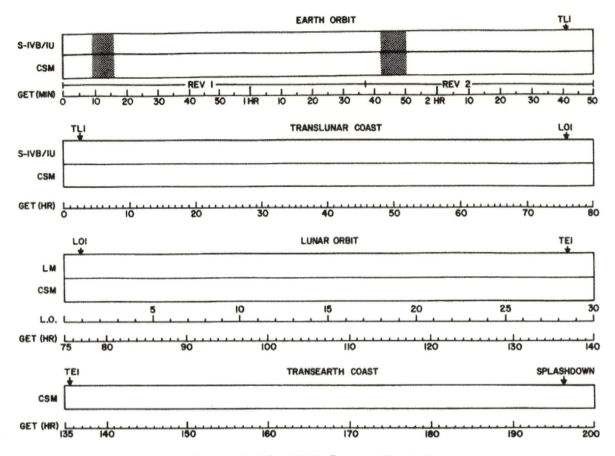

Figure 3-112. VAN Support Periods

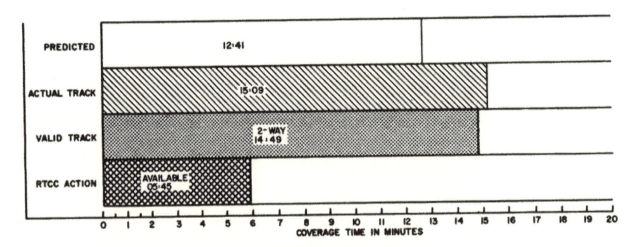

(a) System No. 1, EO -- CSM

Figure 3-113. VAN USB Tracking Coverage

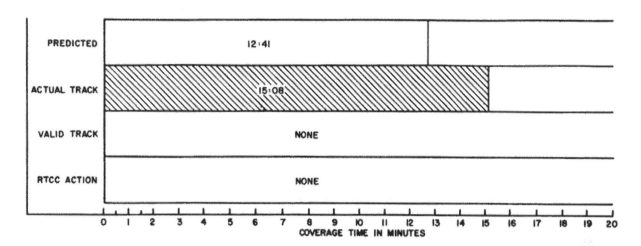

(b) System No. 2, EO -- IU

Figure 3-113. VAN USB Tracking Coverage (cont)

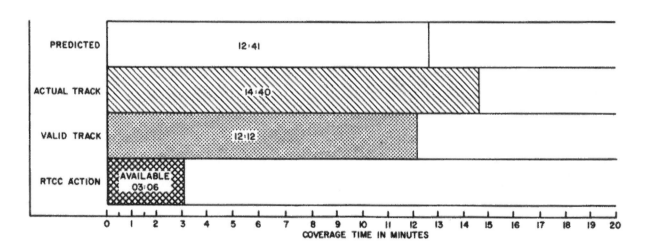

Figure 3-114. VAN FPS-16 Tracking Coverage, EO -- IU

3-131

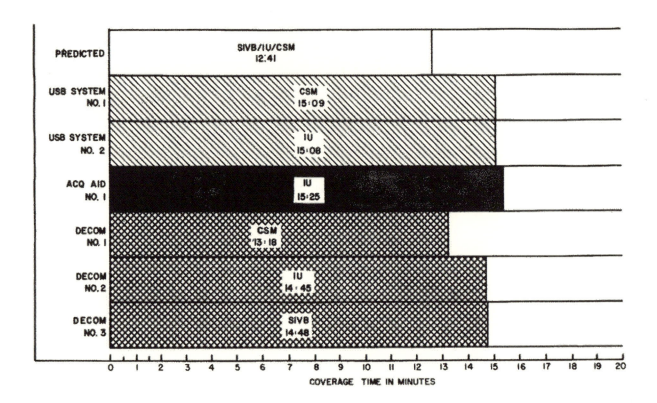

Figure 3-115. VAN Telemetry Coverage, EO

NASCOM NETWORK

PERFORMANCE WITH ANALYSIS

DURING THE APOLLO 11 MISSION

JULY 16-24, 1969

Prepared by

Network Review and Analysis Branch
NASA COMMUNICATIONS DIVISION

Approved by

Joseph G. Sobala

GODDARD SPACE FLIGHT CENTER
Greenbelt, Maryland

Appendix A

SUMMARY

The NASCOM Network satisfactorily supported the APOLLO 11 Mission.

Net 1 was monitored for a total of 52:00 hours. Of the 1,084 transmissions made during this time, 6 were repetitions. The percentage of transmissions not requiring repetitions was 99.45. Voice communications were considered to be excellent.

The Network's high-speed telemetry was monitored for a total of 381:18:45 hours during the mission. Compilation of the data monitored revealed a block error rate of 2.66×10^{-4} and a valid data through-put percentage of 99.97.

Compilation of the Network's high-speed tracking data revealed a block error rate of 2.67×10^{-5} and a data through-put percentage of 99.99.

Monitoring of the Houston wideband primary circuit GW-58526 recorded a total of 21,340,991 blocks received and 2,819 errored blocks, resulting in a block error rate of 1.32×10^{-4} and a valid through-put percentage of 99.99. Monitoring of the secondary circuit GW-58527 recorded a total of 5,244,801 blocks received and 1,282 errored blocks, resulting in a block error rate of 2.44×10^{-4} and a valid through-put percentage of 99.98.

To develop a traffic load study, teletype traffic was surveyed during the mission flight period. A total of 58,977 messages were received and 130,585 were transmitted, developing an input average of 4.55 messages per minute, and an output average of 10.08 messages per minute.

```
                    NASCOM NETWORK
                PERFORMANCE WITH ANALYSIS
              DURING THE APOLLO 11 MISSION


                   July 16-24, 1969


                      CONTENTS
```

SECTION	PAGE
SUMMARY.	A-iii
GENERAL.	A-1
DESCRIPTION OF NETWORK MONITORING.	A-7
PRE-MISSION SYNOPSIS	A-13
MISSION PERFORMANCE.	A-14
SCAMA.	A-14
HIGH SPEED DATA.	A-14
TELETYPE	A-16
POST MISSION REPORT.	A-17
RECOMMENDATIONS.	A-24

```
                      FIGURES
```

NUMBER		PAGE
1	NASCOM NETWORK.	A-3
2	APOLLO 11 NETWORK	A-5
3	VOICE MONITORING EQUIPMENT.	A-9
4	HOUSTON WIDEBAND MONITORING EQUIPMENT	A-10
5	TELEMETRY MONITORING EQUIPMENT.	A-11
6	TRACKING MONITORING EQUIPMENT	A-12

ATTACHMENTS

NUMBER

1   NUMBER OF TRANSMISSIONS MONITORED ON NET 1

2   HOUSTON WIDEBAND BLOCK ERROR RATES GW-58526, GW-58527

3   TELEMETRY DATA BLOCK ERROR RATES

4   TRACKING DATA BLOCK ERROR RATES

5   TELETYPE 494 TRAFFIC LOAD

GENERAL

The NASA Communications Division is the GSFC organization responsible for all operational communication circuits and facilities transmitting mission-related information. This responsibility is fulfilled by a global communication system known as the NASCOM Network.

The NASCOM Network, see Figure 1, page 3, provides the communication links between the Mission Control Center and remote tracking sites through the primary Switching Center at GSFC and five overseas Switching Centers.

For the APOLLO 11 Mission, portions of the network were specifically configured to provide the required communications support. The network configuration for this mission support is shown in Figure 2, page 5. One or more circuits of each net were monitored during the entire mission flight period. Results of this monitoring are presented in this report.

A-1

'A-2

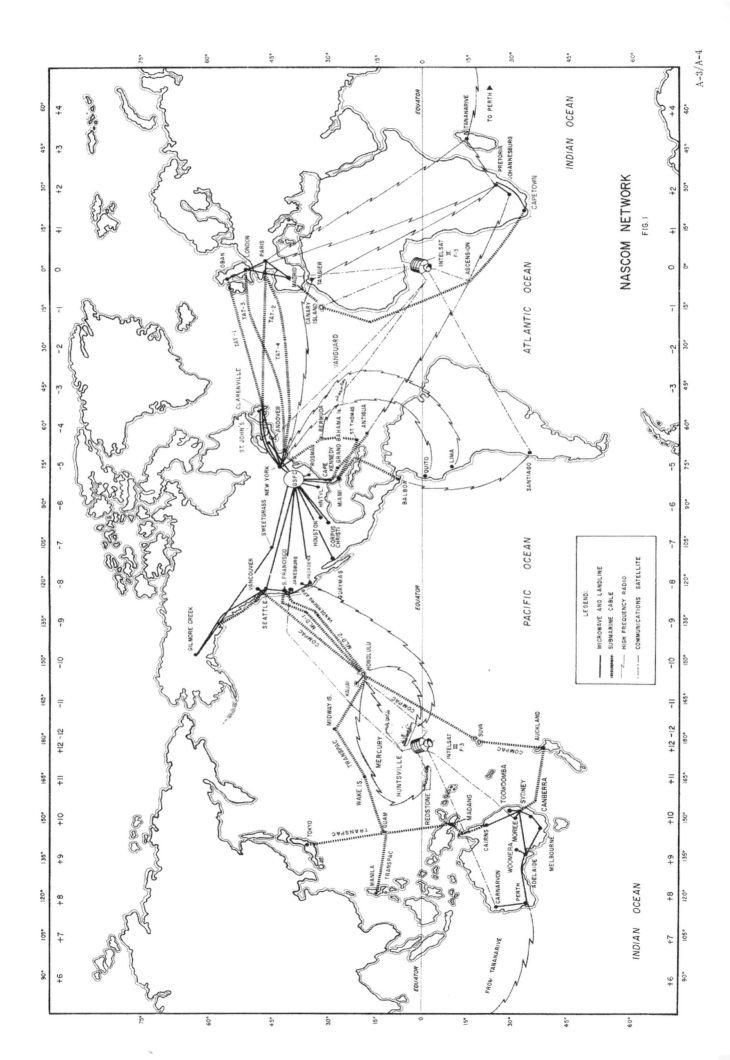
NASCOM NETWORK
FIG. 1

NASCOM

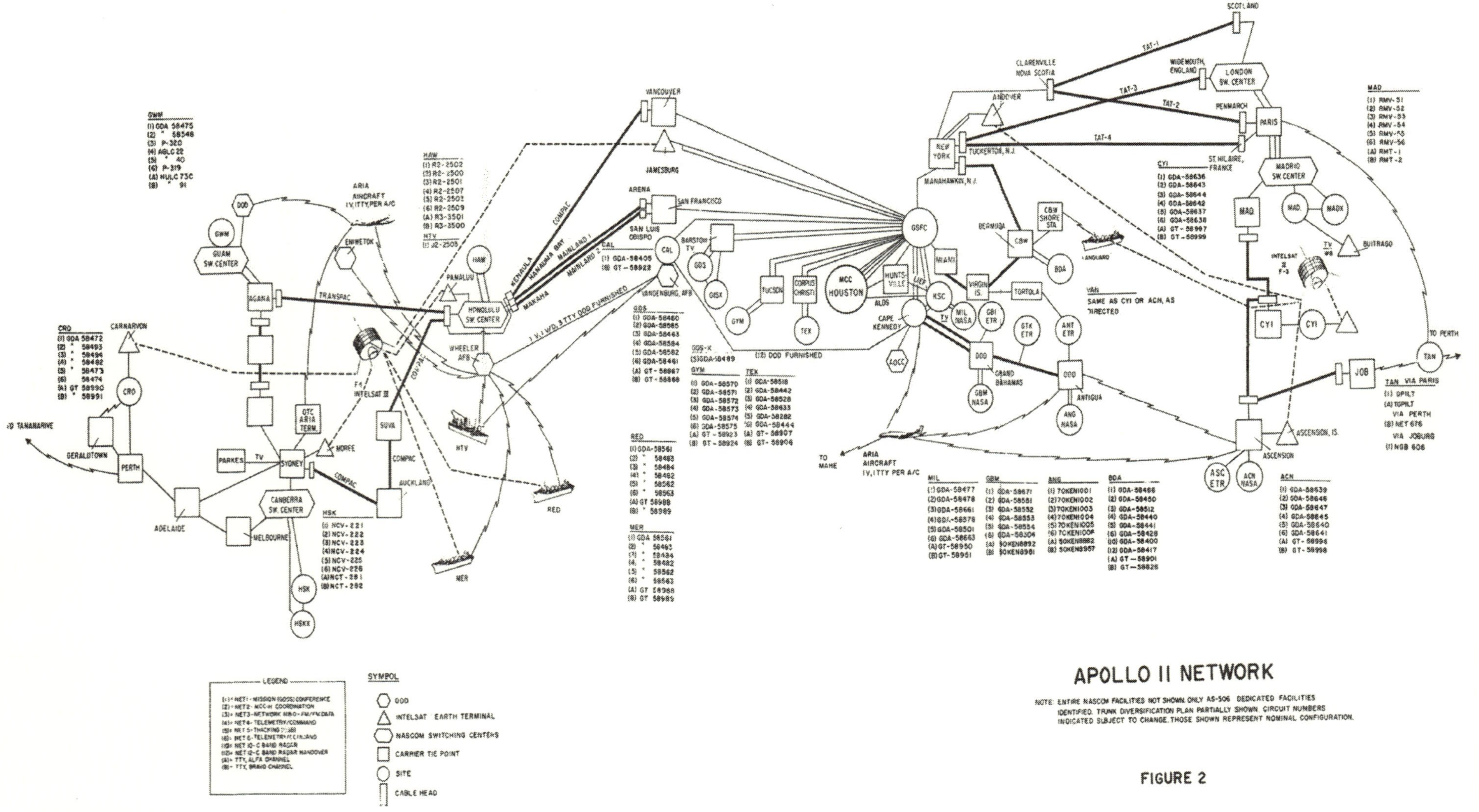

APOLLO II NETWORK

NOTE: ENTIRE NASCOM FACILITIES NOT SHOWN. ONLY AS-506 DEDICATED FACILITIES IDENTIFIED. TRUNK DIVERSIFICATION PLAN PARTIALLY SHOWN. CIRCUIT NUMBERS INDICATED SUBJECT TO CHANGE. THOSE SHOWN REPRESENT NOMINAL CONFIGURATION.

FIGURE 2

PREPARED BY NASA COMM DIV.
COMM OPNS BRANCH, GSFC, 6/25/69

DESCRIPTION OF THE NETWORK MONITORING

Voice monitoring was accomplished on Net 1, as shown in Figure 3, page 10.

The Houston (Wideband) 50.0 kbps primary and secondary data channels from GSFC to MCC-H were monitored to determine the ratio of errored blocks* to total blocks received. The primary and alternate circuits GW-58526 and GW-58527 passed a 50.0 kbps data stream configured in 600 bit message blocks. The test equipment configuration is illustrated in Figure 4, page 11.

Net 4 (Telemetry Data) projected block error rates were determined by: Using the DQM-1 and Model 25-800 Non-Linear System equipment as shown in Figure 5, page 12, and by analyzing the output of telemetry monitored in real time by the Univac 418 with the system at GSFC configured as in Figure 5, page 12. The 418 output provided an octal print-out of the first 20 eight-bit characters and the last eight-bit character which comprise the subframe ID counter of the data stream. To provide this data, errored messages were flagged and the sync group inspected for errors.

Projected block through-put per cent was derived by subtracting errored blocks from total blocks transmitted, dividing this difference by total blocks transmitted, and the quotient converted to per cent. The telemetry data transmission rate was 2.4 kbps with 2400 bit block message lengths.

A-7

Net 5 (Tracking Data). The tracking data circuits were monitored using the Model 547 Error Detection Decoder equipped with suitable counters and recorders as illustrated in Figure 6, page 13. The total number of blocks received for each circuit tested was derived by dividing the data transmission bit rate by the block bit length and multiplying this quotient by the total monitored time in seconds. Actual block errors* were recorded by counts provided by the in-line decoders.

* The actual number of errored blocks, recorded during the monitoring of the Houston Wideband channels and the Net 5 circuits, was supplied by the appropriate in-line decoder. A code unit of 33 bits for each message block is added to the message when transmitted by a TDP. This "TAG" code represents the remainder of a polynomial division performed on the header and message bits. The header, message and code bits then form a complete block of data. The decoder at the receive end re-divides this block of data. This process yields a quotient with no remainder so long as the receive bits are error free. Therefore, this zero remainder identifies a valid block of data. The block error counts for the Houston channels were logged from a digital counter display. The error counts for Net 5 were similarly totaled, but the count was machine printed at one-second intervals.

A-8

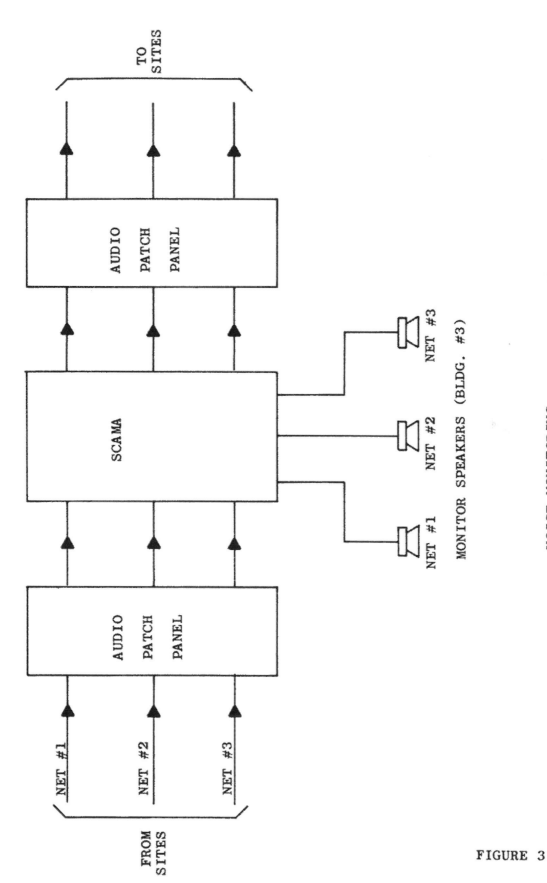

FIGURE 3

VOICE MONITORING
EQUIPMENT CONFIGURATION

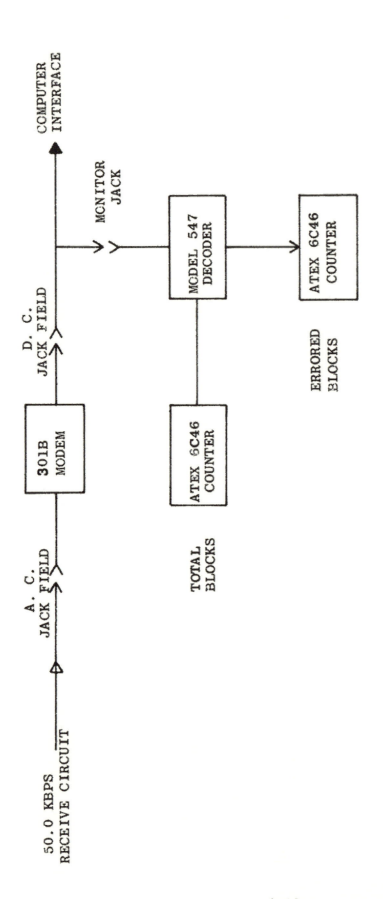

MONITOR EQUIPMENT CONFIGURATION
HOUSTON WIDEBAND DATA SYSTEM

FIGURE 4

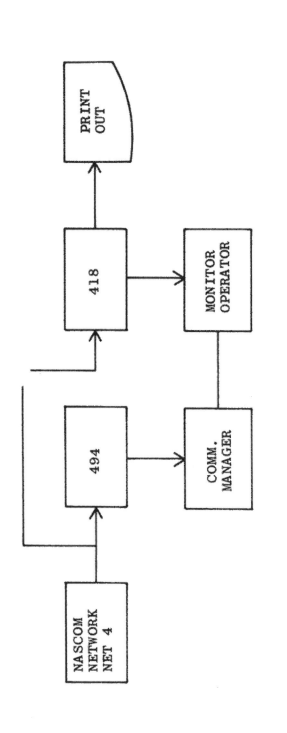
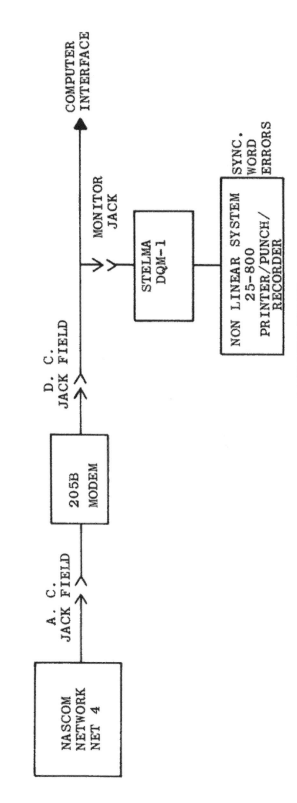

MONITOR EQUIPMENT CONFIGURATION
TELEMETRY DATA

FIGURE 5

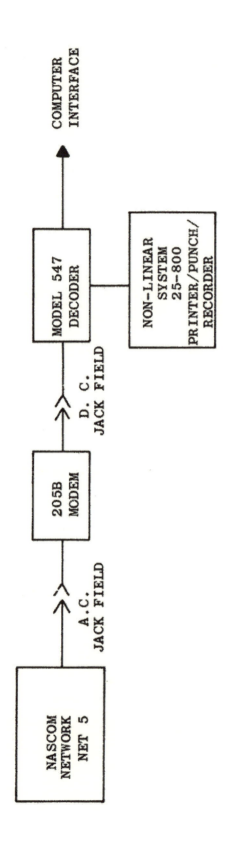

MONITOR EQUIPMENT CONFIGURATION
TRACKING DATA

FIGURE 6

PRE-MISSION SYNOPSIS

NCG 044/725 CDDT/NET SIM

NASCOM support commenced at 02/0430Z July, 1969. Because of the loss of the INTELSAT III, F2 Satellite, circuit facilities into the Madrid area were limited. The normal mission configuration could not be provided to the LMAD station necessitating rescheduling of LMAD in the pre-count for their confidence testing. Circuits were established via LCYI to provide Madrid support and included conferencing LMAD NET 2 with LCYI plus patching two circuits through to GSFC for the LMAD NETS 4 and 6. This arrangement remained in effect until 0723Z when circuits were available between the London and Madrid Switching Centers to establish normal circuit routes to LMAD. At 0645Z, the MCC reported they were unable to transmit high speed commands to the PHAW station. It was determined the problem was caused by an error at GSFC/NASCOM in selecting the long-lines portion of the PHAW Net 4 as a make good for another facility. As soon as this was detected, the circuit was restored to service at 0648Z. A simulated failure of the "A" Voice Frequency Telegraph (VFTG) circuit between GSFC and the MCC was made at 1523Z, to determine personnel responsiveness and speed of restoration.

MCG 725E/NET SIM 7/11/69

There were no significant events which effected support.

A-13

MISSION PERFORMANCE

SCAMA

Monitoring of Voice Net 1 was performed at GSFC from lift-off of
the APOLLO 11 vehicle at 1332Z on July 16th, through the next 7 hours,
for 8 hours each day on July 17th, 18th, 21st, 22nd, and 23rd, and for
5 hours prior to splashdown at 1649Z on July 24th.

Net 1 performed with consistent excellence. During the 52:00 hours
monitored, there were 6 transmissions. The percentage of transmissions
not requiring repetition was 99.45. Tabulated results are presented in
Attachment 1.

HIGH SPEED DATA

The APOLLO 11 Mission support communications components monitored
were as follows: Houston wideband (GW-58526, GW-58527); Nets 4 and 6
telemetry, and Net 5 tracking.

A. Wideband Data: Refer to Attachment 2 for the recorded data results.

The Houston GW-58526 wideband 50.0 kbps circuit was the primary circuit
during the entire mission. Counts of 21,340,991 total blocks and 2,819
errored blocks were recorded. For this circuit the derived figures, for
block error rate and valid through-put percentage, were 1.32×10^{-4} and 99.99,
respectively.

GW-58526 was not monitored during the following periods for the following reasons. On July 17th, the circuit was out from 1820Z to 2019Z because
of a defective modem at Houston. This same trouble reoccurred on July 18th

from 0252Z to 0305Z. Also on the 18th, from 1629Z to 1655Z, outage was caused by carrier equipment trouble. On July 20th, outages occurred between 0422Z and 0530Z because of carrier equipment failure, between 2103Z and 2115Z because of microwave failure on the Garden City/Leesburg link, and between 2155Z and 2158Z because of carrier failure.

The Houston GW-58527 wideband 50.0 kbps circuit was used as an overflow circuit for the primary channel GW-58526. Totals of 5,244,801 blocks received and 1,282 errored blocks were recorded, resulting in a block error rate of 2.44×10^{-4} and a valid through-put percentage of 99.98.

On July 17th, GW-58527 was not monitored from 0545Z to 0648Z because of microwave failure between Jasper, Alabama, and Houston, Texas.

B. <u>Telemetry Data</u>: Refer to page 7 for computation criteria. Attachment 3 shows that 12 telemetry stations were monitored for periods of from 04:30 minutes (Vanguard GDA-58642/GDA-58638), to 126:37:16 hours (Honeysuckle NCV-224/NCV-226). Of the 12 stations monitored, 4 were error free. Mercury had the poorest block error rate of 1.34×10^{-3} and a maximum through-put percentage of 99.87.

A Total of 1,372,125 blocks was transmitted with 365 blocks in error. This yields a mission computed telemetry block error rate of 2.66×10^{-4} and a through-put percentage of 99.97.

C. <u>Tracking Data</u>: Refer to page 8 for computation criteria. Attachment 4 shows that Net 5 tracking stations were monitored for periods of from 05:30 minutes (Bermuda GDA-58441), to 3:05:16 hours (Carnarvon GDA-58473).

A-15

Of the 6 stations monitored, 4 were error free. Bermuda had the poorest block error rate of 3.03×10^{-4} and a maximum through-put percentage of 99.97.

A total of 262,020 blocks was transmitted with 7 blocks in error. This yields a mission computed tracking block error rate of 2.67×10^{-5}, and a through-put percentage of 99.99.

TELETYPE

Totals of daily Houston IN and OUT messages were provided by the 494 Switching Computer (see Attachment 5). IN messages ranged from a low of 5,868 on July 24th to a high of 7,962 on July 16th, with a low average per minute of 4.08 messages and a high average per minute of 5.53. OUT messages ranged from a low of 12,738 on July 19th to a high of 17,447 on July 16th, with a low average per minute of 8.85 messages, and a high average per minute of 12.12. There was a total of 58,977 IN messages and a total of 130,585 OUT messages, for a grand total of 189,562, developing an overall average per minute of 7.31 messages.

In comparison with the previous mission, IN messages decreased by .22 per cent per minute, and OUT messages decreased by 1.09 per cent per minute.

POST MISSION REPORT

NASCOM support was considered satisfactory for the mission. Few major items occurred that affected support; those items considered significant are discussed below.

15/0200Z. The MCC TIC reported data dropouts. The wideband "A" (GW-58526) was set out and FE600/TMG tests run. The MCC shifted to the "B" data circuit from the "A" circuit. The "A" data circuit was released to AT&T. The problem of data dropouts on the "A" circuit continued until 15/2009Z when it was determined crosstalk caused by the ringers of the voice orderwires on the wideband were causing the problem. A procedure to have the facility controllers use the GOSS 8 orderwire was instituted and the modem orderwires were disabled.

16/0645Z. The MVAN reported his downlink via SATCOM was noisy. This created a problem running the command interface with the MCC on time. Power on the Intelsat satellite was increased 1 dB which made a slight improvement. Another 1 dB increase caused the VANGUARD to interfere with the MADRID Wideband because of intermodulation products in the satellite. Between 0645Z and approximately 0845Z various 2.4 kbs checks were made via SATCOM and HF. As a result of these checks and after the necessary coordination, the transmit to the MVAN was dropped to 1.2 kbs and the receive remained at 2.4 kbs. Ths system operated satisfactorily in this manner.

A-17

16/1446Z. No low speed tracking data was received from LTAN. A check of monitors at the Bassendean, Australia receiving station and the CANBERRA switch showed no circuit problems. No reason can be given for lack of data since the circuits were apparently good.

16/1612Z. The MERCURY was brought up in the CARNARVON satellite frequency slot, however the high speed (NET 5) tracking data kept starting and stopping even though the circuit was good. It appeared that the MERCURY had a computer problem since their computer went down at 1620Z.

17/0158Z. After receiving a report of excessive errors on the GSFC/MCC wideband "B" (GW-58527) circuit it was set out from both computers. The problem was isolated between Falkner and Waldorf, Maryland. The circuit was returned to service at 0206Z.

17/0430Z. The following is a quote of GSTS 17/1824Z.

17/0430Z. MCC TIC experiencing TLM dropouts on GW-58526. MCC COMM Control detected errors on decoder also.

0432Z. GW-58526 came clear of errors.

0447Z. GW-58526 out transmit; high error count at COMM Control.

0459Z. GW-58526 returned to carrier for corrective action.

0512Z. GW-58526 put on make-good circuit and tested 5/5; put back in service.

0514Z. GW-58526 MCC again seeing errors as soon as circuit put back to data.

A-18

0545Z.   GW-58527 out receive.

0600Z.   GW-58527 returned to carrier.

0610Z.   GW-58526 switched modems at GSC. Data still no good to MCC.

0613Z.   GW-58526 normalled up modems. Master cleared pbt's and restarted output. Data still no good to MCC.

0616Z.   Switched entire CP system at GSC. Data still no good to MCC.

0639Z.   GW-58527 in both ways. Outage due to MUR-201 failure Houston/Jasper, Alabama.

0652Z.   GW-58526 restored to normal facilities and operated good end-to-end. Carrier reports there definitely was trouble on the circuit but came good before it could be located.

SUMMARY: Both GW-58526 and GW-58527 experienced valid carrier problems during the time period covered. The spare circuit utilized to replace GW-58526 during the failure would pass test data, FACS control to FACS control but would not pass data CP to CP. The reason for this is undetermined. At this time we are in verbal contact with COMM control regarding further attempts to run additional tests with the same configuration as when the trouble was experienced.

NOTE: Extensive tests run the next day and through to 19/1316Z including swapping the "A" and "B" circuits with the overhead (TPL)

A-19

spare, running various tests, etc., resulted in the replacement of the "A" data set at the MCC with a new one. On the 20th, at 1150Z, the "B" circuit was acceptable for data transfer on either data set after Southwestern Bell replaced a "roll off filter card" in the "B" data set at the MCC. Part of the initial problem of restoration of the "B" circuit on the 17th was having the MCC go into a "forced mode" which is an indication of a circuit problem. Further testing revealed that when the 494 interface fails over to the alternate circuit, the MCC ceases to send test blocks and goes into a "forced mode" unless 5 consecutive acknowledgements are received. Further testing determined the only way the MCC can come out of this "forced mode" is to send at least 5 commands in order to receive 5 consecutive acknowledgements in order to set the circuit normal. We recommended further testing be done to determine the specific problem and what must be done to correct it.

17/0545Z. GW-58527 GSFC/MCC out. At 0648Z the circuit was restored to service. Problem was an MUR-1 failure between Jasper, Alabama and Houston, Texas.

17/2004Z. The following is a quote of GSTS 17/2335Z:

"Reference CP problem at GSFC from 2004Z to 2041Z. The change made to eliminate the delay experienced in tracking data during previous Apollo missions was to assign a higher drum priority to the 8 low speed data circuits. When the CP started receiving 1 sample each 10

A-20

seconds, this change effectively caused the 8 low speed circuits to block all other output routines from the drum during the 4 seconds out of each 10 when the input tracking circuits were idle. This also accounted for the MADRID CP problems and the sporadic output and validation build up. The problem was corrected by making a core change to put the 8 circuits back to their original priority. This change is valid as long as 1 sample per 10 seconds is being received. If track goes back to 1 sample per 6 seconds a core change will have to be made back to the higher priority. Advance notification to the COMMGR is mandatory if a sample every 10 seconds will be required."

18/1250Z. It was noted that low speed data transmission from LMAD and the wing was disrupted going towards the MCC. It was determined that the 418 switching computer at the MADRID Switching Center and the 494 program at the GSFC had an anomaly which prevented the retransmission of a segment of data with a message number above 177 (octal). In order to preclude against data stoppage, LMAD and LKMW were extended directly to GSFC for input to the 494.

18/2138Z. GBDA low speed tracking data garbling towards the MCC. The data looked good on input at GSFC. GSFC facilities switched the circuit to channel 2 of the VFTG and the data was good at 2144Z. The problem was determined to be a bad channel in the GSFC VFTG System B.

19/0700Z. A coordinated switch of the CP's was made due to reports of possible thunderstorms in the Goddard area. The "B" computer on diesel

A-21

became the primary on-line system.

22/1137Z. GW-58526 GSFC/MCC wideband "A" circuit began showing errors at the MCC. MCC set out the transmit to GSFC which forced the data onto the alternate, GW-58527, circuit to MCC. The "A" wideband circuit was made good on the spare circuit at 1147Z. The normal circuit came good at 1151Z, however it was not restored to service per coordination COMMGR/COMM Control.

22/1219Z. High Speed output from GSFC to the MCC on the "A" wideband failed over to the "B", however, the MCC reported not seeing any data on the "B" circuit. At 1224Z, the MCC reported the problem was because of a patching error on the PBT's at the MCC.

22/2350Z. The "B" 58527 circuit GSFC/MCC went to a test mode. The MCC reported not seeing any data or test on the circuit. At 23/0001Z, the "B" wideband circuit was made good with the normal "A" wideband, GW-58526. The problem was a result of a microwave failure between GSFC and Suitland, Maryland, because of a severe thunder and rain storm.

23/0004Z. A power dip due to the storm caused a fault of the off-line CP, which also caused the console to hang up, preventing system initialization. The "C" mainframe was brought up with the "B" peripherals at 0014Z.

23/0600Z. After coordinating with COMM Control, a change was made to restore the "A" computer as the prime "off-line" system to maintain the diesel/commercial power configuration integrity.

23/0709Z. The "A" system was placed on line in place of the "B" with a fresh program load. This was coordinated with COMM Control prior to the change. As a point of interest, a check with TIC noted a loss of approximately one second of high speed TLM during the change.

23/0814Z. Both wideband systems normalled at this time.

23/0830Z. After a request by the ALSEP Net Control at 0330Z, extensive checks were made on the TEXAS ALSEP high speed telemetry input to determine if a problem was evident with the frame sync and message/segment numbers. Numerous delogs were taken here and no problems could be found.

23/1610Z. The "B" system was placed on-line to provide diesel power rather than commercial.

No other problems of note were experienced for the duration of the mission.

ATS support. The ATS VHF capability was used with excellent success from the prime recovery ship USS Hornet. During recovery, the circuit was used exclusively by the White House representative on the Hornet to coordinate the President's schedule with the USS Arlington, Pacific Radio, Johnston Island, the USS Hornet and the White House.

A-23

RECOMMENDATIONS

1. On the 18th through the 22nd radio day, numerous messages were skipped after a queue count showed more than ten messages on queue for GQMI, GQNA, GQKC, and GQDA. The NASCOM Division believes that the transmissions to these stations should be reviewed between now and APOLLO 12 and reduced as much as possible. The following count shows the stations and numbers of messages skipped.

	GQMI	GQNA	GQKC	GQGA
July 18	21	127	127	-
19	35	177	177	-
20	147	423	641	430
21	170	469	731	536
22	163	461	731	568

2. Reference procedures for MCC/GSFC operations interface procedures. Recommend the procedures transmitted prior to APOLLO 11 relating to circuit reconfigurations be coordinated only between the COMM Control/ COMMGR. As an example, at approximately 24/0658Z, TIC network was configuring circuits to ask for 3 telemetry streams. This involves NASCOM action at GSFC, that is, changing the CP input from a polynomial terminal to a CLT. It is imperative that one common circuit configuration control point be established and that must be between the COMM Control and COMMGR., not the end users and the station involved.

A-24

MONITORING OF

VOICE TRANSMISSIONS/RETRANSMISSIONS ON NET-1

A	B		C	D	E	F	G
DATE	TIME GMT		△ T HR/MIN	ORIGINAL XMSNS	RE-XMSNS	TOTAL XMSNS (D+E)	PERCENT (D÷F)
	START	STOP					
7/16/69	1330	2030	7:00	288	0	288	100.00
7/17/69	1200	2000	8:00	283	0	283	100.00
7/18/69	1200	2000	8:00	103	1	104	99.04
7/21/69	1200	2000	8:00	176	2	178	98.88
7/22/69	1200	2000	8:00	49	3	52	94.23
7/23/69	1200	2000	8:00	72	0	72	100.00
7/24/69	1200	1700	5:00	107	0	107	100.00
TOTALS			52:00	1,078	6	1,084	*99.45

*AVERAGE

ATTACHMENT 1

HOUSTON WIDEBAND

GW-58526

DATE	$\triangle$ T HR/MIN	TOTAL BLOCKS	TOTAL ERRORED BLOCKS	BLOCK ERROR RATE	VALID THRU-PUT PERCENT
7/16/69	10:30	1,229,091	287	2.34×10^{-4}	99.98
7/17/69	21:00	2,499,805	446	1.78×10^{-4}	99.98
7/18/69	16:00	1,698,846	48	2.83×10^{-5}	99.99
7/19/69	19:00	2,201,824	42	1.91×10^{-5}	99.99
7/20/69	17:30	1,891,825	451	2.38×10^{-4}	99.98
7/21/69	24:00	2,912,990	239	8.20×10^{-5}	99.99
7/22/69	24:00	3,291,064	558	1.70×10^{-4}	99.98
7/23/69	24:00	3,250,685	390	1.20×10^{-4}	99.99
7/24/69	17:00	2,364,861	358	1.51×10^{-4}	99.98
TOTALS	173:00	21,340,991	2,819	$*1.32 \times 10^{-4}$	*99.99

*AVERAGE

ATTACHMENT 2-1

HOUSTON WIDEBAND

GW-58527

DATE	$\triangle$ T HR/MIN	TOTAL BLOCKS	TOTAL ERRORED BLOCKS	BLOCK ERROR RATE	VALID THRU-PUT PERCENT
7/16/69	10:30	86,845	15	1.73×10^{-4}	99.98
7/17/69	23:00	540,427	260	4.81×10^{-4}	99.95
7/18/69	23:30	1,357,981	677	4.99×10^{-4}	99.95
7/19/69	24:00	1,064,677	129	1.21×10^{-4}	99.99
7/20/69	24:00	919,892	92	1.00×10^{-4}	99.99
7/21/69	24:00	469,309	29	6.18×10^{-5}	99.99
7/22/69	24:00	299,905	38	1.27×10^{-4}	99.99
7/23/69	24:00	311,251	24	7.71×10^{-5}	99.99
7/24/69	17:00	194,514	18	9.25×10^{-5}	99.99
TOTALS	194:00	5,244,801	1,282	$*2.44 \times 10^{-4}$	*99.98

*AVERAGE

ATTACHMENT 2-2

MONITORED BLOCK ERROR RATES

NET-4 & 6 (TELEMETRY DATA)

STATION	CIRCUIT NET 4	NET 6	$\triangle$ T: HOURS: MINUTES: SECONDS	TOTAL BLOCKS	TOT. ERR. SYNC WDS.	BLOCK ERROR RATE	MAXIMUM THRU-PUT PERCENT
ASCEN-SION	GDA-58645	GDA-58641	01:00:18	3,618	0	0.00×10^{-0}	100.00
CANARY ISLAND	GDA-58642	GDA-58638	04:04:50	14,690	10	6.81×10^{-4}	99.93
CAR-NARVON	GDA-58482	GDA-58474	01:09:14	4,154	1	2.41×10^{-4}	99.98
GOLD-STONE	GDA-58584	GDA-58461	86:55:18	312,918	81	2.59×10^{-4}	99.97
GUAM	AGLC-22	P-319	14:04:12	50,652	6	1.18×10^{-4}	99.99
HAWAII	R2-2507	R2-2509	18:46:55	67,615	12	1.77×10^{-4}	99.98
HONEY-SUCKLE	NCV-224	NCV-226	126:37:16	455,836	58	1.27×10^{-4}	99.99
MADRID	RMV-54	RMV-56	125:41:49	452,509	194	4.29×10^{-4}	99.96
MER-CURY	GDA-58482	GDA-58563	00:37:22	2,242	3	1.34×10^{-3}	99.87
MERRITT ISLAND	GDA-58578	GDA-58663	00:38:23	2,303	0	0.00×10^{-0}	100.00
RED-STONE	GDA-58482	GDA-58563	01:38:38	5,318	0	0.00×10^{-0}	100.00
VAN-GUARD	GDA-58642	GDA-58638	00:04:30	270	0	0.00×10^{-0}	100.00
TOTALS			381:18:45	1,372,125	365	$*2.66 \times 10^{-4}$	*99.97

*AVERAGE

ATTACHMENT 3

MONITORED BLOCK ERROR RATES

NET-5 (TRACKING DATA)

STATION	CIRCUIT	Δ T HOURS: MINUTES: SECONDS	TOTAL BLOCKS	TOT. ERR. SYNC. WDS.	BLOCK ERROR RATE	MAXIMUM THRU-PUT PERCENT
HAWAII	R2-2503	00:53:37	32,170	6	1.86×10^{-4}	99.98
BERMUDA	GDA-58441	00:05:30	3,300	1	3.03×10^{-4}	99.97
MADRID	RMV-55	01:08:42	41,220	0	0.00×10^{-0}	100.00
GOLD-STONE	GDA-58582	01:10:06	42,060	0	0.00×10^{-0}	100.00
CANARY	GDA-58637	00:53:31	32,110	0	0.00×10^{-0}	100.00
CARNARVON	GDA-58473	03:05:16	111,160	0	0.00×10^{-0}	100.00
TOTALS		07:16:42	262,020	7	$*2.67 \times 10^{-5}$	*99.99

*AVERAGE

ATTACHMENT 4

TELETYPE

494 TRAFFIC LOAD

DATE	MESSAGES IN	AVERAGE IN PER MINUTE	MESSAGES OUT	AVERAGE OUT PER MINUTE
7/16/69	7,962	5.53	17,447	12.12
7/17/69	6,527	4.53	14,216	9.87
7/18/69	6,737	4.68	14,921	10.36
7/19/69	6,060	4.21	12,738	8.85
7/20/69	5,998	4.17	13,368	9.28
7/21/69	6,186	4.30	14,283	9.92
7/22/69	7,065	4.91	15,955	11.08
7/23/69	6,574	4.57	14,457	10.04
7/24/69	5,868	4.08	13,200	9.17
TOTALS	58,977	4.55	130,585	10.08

ATTACHMENT 5

APPENDIX B – List of Abbreviations and Acronyms

ac	Alternating current
ACN	Ascension Island MSFN station
acq aid	Acquisition aid
AFC	Automatic frequency control
AFETR	Air Force Eastern Test Range
AFWTR	Air Force Western Test Range
A/G	Air-to-ground
AGAVE	Automatic gimballed antenna vector equipment
AGC	Automatic gain control
AIS	Apollo instrumentation ship
ALSEP	Apollo lunar surface experiment package
AM	Amplitude modulation
ANG	Antigua Island MSFN station
AOCC	Aircraft Operations Control Center
AOS	Acquisition of signal
APB	American Projects Branch of the Australian DOS
APP	Antenna position programmer
APS	Ascent propulsion system
ARIA	Apollo range instrumentation aircraft
ATS	Applications technology satellite
AZ	Azimuth
BCD	Binary coded decimal
BDA	Bermuda MSFN station
BER	Bit error rate
biomed	Biological-medical
CADCPS	Combined antenna drive command program system
CADFISS	Computation and data flow integrated subsystem
CAL	South Vandenberg, California C-band station
CAM	Computer address matrix
CAP	Command analysis pattern, also command acceptance pulse

B-1

CapCom	Capsule communicator
CAPRI	Compact all-purpose range instrument
C-band	3900 to 6200 MHz
CBARF	Command backup automatic recovery feature
CCATS	Command communication and telemetry system
CCS	Command control system
CDP	Central data processor
CDR	Commander
CDDT	Countdown demonstration test (prelaunch)
CF	Center frequency
CM	Command module
CMD	Command
CMP	Command module pilot
ComTech	Communications technician
CP	Communications processor
CRF	Capsule radiation frequency
CRO	Carnarvon, Australia MSFN station
CRR	Change recommendation report
CSC	Contingency sample collection
CSM	Command service module
CW	Continuous wave
CYI	Canary Island MSFN station
DAC	Digital-to-analog converter
dB	Decibel
dBm	Decibel referred to 1 milliwatt
dc	Direct current
DCI	Documentation change instruction
DCN	Documentation change notice
DDF	Digital data formatter
DDMS	Department of Defense Manned Flight Support Office

decom	Decommutator
demod	Demodulator
DOD	Department of Defense
DOI	Descent orbit insertion
DOS	Australian Department of Supply
DPS	Descent propulsion system
DSDU	Decommutation system distribution unit
DSE	Data storage equipment
DSIF	Deep space instrumentation facility
DSN	Deep Space Network
DSS	Data Services Section
DSS-71	Deep Space Station-71 (JPL, Merritt Island)
DTU	Data transmission unit
EASEP	Early Apollo surface experiment package
EI	Engineering instruction
EKG	Electrocardiogram
EL	Elevation
EMOD	Erasable memory octal dump
EMU	Expanded memory unit
EO	Earth orbit
EOF	End of file
EOI	Earth orbit insertion
ETA	Estimated time of arrival
ETO	Estimated time of operation
EVA	Extra vehicular activity
FLTLD	Fault load tape (automatic recovery program)
FM	Frequency modulation
fps	Feet per second
FSR	Flight support request

GBM	Grand Bahama Island MSFN station
GCC	Ground communication coordinator
GDS	Goldstone, California MSFN prime station
GDSX	Goldstone, California wing station
GET	Ground elapsed time
GMT	Greenwich mean time
GOSS	Ground operations support system
GRTC	Goddard real-time computer
GRTS	Goddard real-time system
GSFC	Goddard Space Flight Center, Greenbelt, Maryland
GWM	Guam, Marianas Island MSFN station
GYM	Guaymas, Mexico MSFN station
HAW	Kokee Park, Kauai Island, Hawaii MSFN station
HBR	High bit rate
HBRD	High bit rate dump
HF	High frequency (3 to 30 MHz)
HGA	High-gain antenna
HS	High speed
HSD	High-speed data
HSK	Honeysuckle Creek, Australia MSFN prime station
HSKX	Honeysuckle Creek, Australia wing station
HSP	High-speed printer
HTV	USNS Hunstville
Hz	Hertz (cycle/sec)
IC	Instrumentation coordinator
IDRAN	Integrated digital range
IF	Intermediate frequency
I/O	Input/output
IRIG	Interrange instrumentation group (telemetry fequency)
IRV	Interrange vector

ISA	Interface systems adapter
ISI	Instrumentation support instruction
IST	Integrated system test
IU	Instrumentation unit
JPL	Jet Propulsion Laboratory, Pasadena, Calif.
kbps	Kilobits per second
kbs	Kilobits
kHz	Kilohertz
KSC	Kennedy Space Center, Cape Kennedy, Florida
LBR	Low bit rate
lbs	Pounds
LDN	London
LES	Launch escape system
LM	Lunar module
LMP	Lunar module pilot
LO	Lunar orbit
LOI	Lunar orbit insertion
Loran	Long-range navigation
LOS	Loss of signal
LS	Low speed
LSB	Least significant bit
LSD	Low-speed data
LSR	Launch support request
LTDS	Launch trajectory data system
LV	Launch vehicle
LVDC	Launch vehicle digital computer
ma	Milliampere
MAD	Madrid, Spain MSFN prime station
MADX	Madrid, Spain wing station
MAP	Message acceptance pulse

MARS	Goldstone DSN 210-foot antenna
MCC	Mission Control Center Houston, Texas
MER	USNS Mercury
MFED	Manned Flight Engineering Division
MFOD	Manned Flight Operations Division
MFPAD	Manned Flight Planning and Analysis Division
MHz	Megahertz
MIL	Merritt Island, Florida MSFN station
mm	Millemeter
MMR	M&O postmission report
M&O	Maintenance and operations
MOCR	Mission Operations Control Room
modem	Modulator/demodulator
MOM	Maintenance Operations Manual
MSC	Manned Spacecraft Center, Houston, Texas
MSFC	Marshall Space Flight Center, Huntsville, Ala.
MSFN	Manned Space Flight Network
MSFNOC	Manned Space Flight Network Operations Center
MSFTP	Manned space flight telemetry processor
MTU	Magnetic tape unit
NASA	National Aeronautics and Space Administration
NASCOM	NASA Communications Network
NC	Network Controller
NCG	Network Control Group
ND	Network Director
nmi	Nautical mile
NOD	Network Operations Directive
NOM	Network Operations Manager
NRT	Network readiness test
NST	Network support team

NTTF	Network Test and Training Facility at GSFC
ODOP	Offset Doppler velocity and position
omni	Omni-directional antenna
OMSF	Office of Manned Space Flight
OPN	Operations message
OUCH	Offline Universal Command History 642B program
PA	Power amplifier
PAM	Pulse amplitude modulation
PAOS	Predicted acquisition of signal
paramp	Parametric amplifier
PBI	Push button indicator
PBR	Premission briefing report
PCA	Point of closest approach
PC card	Printed circuit card
PCM	Pulse code modulation
PDI	Powered descent initiation
PFS	Precision frequency source
PLIM	Postlaunch instrumentation message
PLSS	Portable life support system
PM	Phase modulation
PMR	Postmission report
preamp	Preamplifier
PRN	Pseudo random noise
PSK	Phase-shift keyed
psi	Pounds per square inch
PSRD	Program support requirements document
PTC	Passive thermal control
RCS	Reaction control system
R/E	Receiver/exciter

RED	USNS Redstone
RF	Radio frequency
RFI	Radio frequency interference
RIC	Request for instrumentation clarification
RSCC	Remote station command computer
RSDP	Remote site data processor
RSO	Range safety officer
RTC	Real-time command
RTCC	Real-time computer complex
RTCS	Real-time computer system
SATCOM	Satellite communications
S-band	1500 to 3900 MHz
S/C	Spacecraft
SCAMA	Switching, conference, and monitoring arrangement
SCM	Site configuration message
SCO	Subcarrier oscillator
SD	Standard deviation
SDDS	Signal data demodulator system
S-IC	Saturn V first stage
S-II	Saturn V second stage
S-IVB	Saturn V third stage
SLA	Spacecraft lunar adapter
SM	Service module
S/N	Signal/noise
SPAN	Solar Particle Alert Network
SPE	Static phase error
SPS	Service propulsion system
SRT	Station readiness test
SST	Subsystem test
ST	System test

STADAN	Space Tracking and Data Acquisition Network
TAN	Tananarive, Malagasy Republic, STADAN station
TDP	Tracking data processor
TEC	Transearth coast
TEI	Transearth injection
TELTRAC	Telemetry tracking (Canoga acq aid system)
TETR-B	Test and training satellite-B
TEX	Corpus Christi, Texas MSFN station
TIC	Telemetry instrumentation coordinator
TLC	Translunar coast
TLI	Translunar injection
TLM	Telemetry
TSI	Test support instruction
TSP	Test support position
TTY	Teletype
TV	Television
TWX	Teletype wire exchange message
UDB	Updata buffer
UHF	Ultra-high frequency
USB	Unified S-band
VAN	USNS Vanguard
VCO	Voltage controlled oscillator
VHF	Very-high frequency
VOGAA	Voice-operated gain adjusting amplifier
VR	Voltage regulator
VSWR	Voltage standing wave ratio

DISTRIBUTION LIST

GSFC

J. Lacy	202	1
J. Riddle	252	2
R. Moore	460	1
J. Mengel	500	1
F. Vonbun	550	2
G. Ludwig	560	1
O. Covington	800	1
R. McCaffery	802	5
T. Roberts	810	2
L. Bonney	811	1
V. Gardner	811	1
D. Dalton	812	2
P. Pashby	813	2
P. Brumberg	813	2
W. LaFleur	814	2
H. Wood	820	1
D. Call	820	1
R. Augestein	820	1
W. Pfeiffer	821	1
C. Knox	821	1
E. Ferrick	821	1
A. Dannessa	821	1
D. Graham	822	15
O. Womick	823	2
D. Spintman	823	1
T. Hetzel	824	1
R. Capo	824	1
W. Varson	830	5
J. Shaughnessy	834	5
L. Stelter	840	7
L. Manning	841	1
D. Tuning	841	1
D. Stoehr (KSC)	841	1
W. Rogers	842	1

BFEC

L. Jochen	Columbia	6
W. Hooper	Columbia	1
C. Simpson	CCS	3
T. Begenwald	DSS	2
D. Smith	EES	2
C. Garrison	ESO	3
W. Lauman	NSG	1
D. Rechenbach	ODS	2
P. Leimanis	ODS	1
S. Berger	ODS	2
N. Lindenberg	OSO	1
D. Greene	PSS	1
E. Thomas	SOS	5
M. Daniels	SOS	12
W. Keesey	SSO	1
R. Pickens	Com Div	1

NASA HQ

J. Raleigh	MAS	3
J. Stockwell	TN	2
F. Bryant	TR	1
W. Folsom	TR	3

MSC

L. Croom	FS4	2
J. Mager	FS4	1
A. Larsen	FS23	1
S. Sanborn	NOB	2
T. Sheehan	FS4	4
R. Rose	FA	5
H. Kyle	EE	10

MSFC

A. McNair	PM-MO	10

NASA REPS

W. Easter	MSC	2
P. Smor	WTR	1
J. Speck	GNSO	4

SHIP REP

O. Thiele	VAN	1

MISC

A. Maloy	AFWTR (WTOOP)	3
R. Perry	AFWTR	1
Capt. Baker	ASORF	4
Col. Mask	DDMS	10
Capt. Honeycutt	AFETR	24
P. Goodwin	JPL	3
P. Cahalan	NA-2D	1
American Projects Branch	DOS	4
L. Robinson	Wheeler Range Comm Control	1

SWITCHING CENTERS

CSW	1
GWM	1
HON	2
LDN	1
MAD	1

TRACKING STATIONS

ACN	2
ANG	2
BDA	4
CRO	3
CYI	4

TRACKING STATIONS (cont)

DBA	2
GBM	5
GDS	2
GDSX	1
GWM	5
GYM	6
HAW	5
HSK	5
HSKX	1
MAD	5
MADX	1
MARS	2
MIL	6
NT&TF	6
PARKES	2
TAN	4
TEX	5
VAN	4

APOLLO PROGRAM FLIGHT SUMMARY
(continued from front cover)

Apollo 7
(AS-205)

Oct 11, 1968

First manned Apollo flight--earth orbital. Demonstrated spacecraft, crew, and support element performance during highly successful flight lasting 10 days 20 hours, included eight major propulsion system firings and first live TV from manned spacecraft.
(Schirra, Cunningham, Eisele)

Apollo 8
(AS-503)

Dec 21, 1968

History's first manned flight from earth to another body in solar system. Included 10 revolutions around moon (20-hour period) and safe return to earth, TV and photography of moon and earth by astronauts.
(Borman, Lovell, Anders)

Apollo 9
(AS-504)

March 3, 1969

Third manned Apollo Saturn flight -- 10 day low earth orbit. Evaluated the first manned LM and demonstrated compatibility of CSM and LM to perform operations typical of a lunar mission.
(McDivitt, Scott, Sweickert)

Apollo 10
(AS-505)

May 18, 1969

Fourth manned Apollo Saturn flight -- 8-day lunar orbit. Demonstrated CSM and LM capabilities in space maneuvering, separation, rendezvous, and docking. Preparation for lunar landing mission.
(Stafford, Young, Cernan)

Apollo 11
(AS-506)

July 16, 1969

Accomplishment of moon landing goal. Man's first footsteps on any celestial body besides earth. EVA included gathering of lunar samples.
(Armstrong, Collins, Aldrin)

Made in the USA
Middletown, DE
22 July 2023

35318959R00146